SURVIVAL TECHNIQUES FOR THE PRACTICING ENGINEER

SURVIVAL TECHNIQUES FOR THE PRACTICING ENGINEER

ANTHONY SOFRONAS

WILEY

For general information on our other products and services or for technical support, please contact our Customer Care Department within the United States at (800) 762-2974, outside the United States at (317) 572-3993 or fax (317) 572-4002.

Wiley also publishes its books in a variety of electronic formats. Some content that appears in print may not be available in electronic formats. For more information about Wiley products, visit our web site at www.wiley.com.

Library of Congress Cataloging-in-Publication Data

Names: Sofronas, Anthony, author.
Title: Survival techniques for the practicing engineer / Anthony Sofronas.
Description: Hoboken, New Jersey : John Wiley & Sons, 2016. | Includes
   bibliographical references and index.
Identifiers: LCCN 2016011022| ISBN 9781119250456 (cloth) | ISBN 9781119250487
   (epub) | ISBN 9781119250500 (epdf)
Subjects: LCSH: Engineering–Vocational guidance.
Classification: LCC TA157 .S63174 2016 | DDC 620.0023–dc23 LC record available at
https://lccn.loc.gov/2016011022

Cover image courtesy of Dr. Sofronas

Typeset in 10/12pt TimesLTStd by SPi Global, Chennai, India

Printed in the United States of America

10 9 8 7 6 5 4 3 2 1

To My Lord Who Has Made This All Possible
And
To My Family And Friends Who Have Contributed To My Success

CONTENTS

ABOUT THE AUTHOR

Anthony Sofronas is an author, educator, and consultant. He has extensive practical experience in troubleshooting machinery and equipment. His 48 years in industry has been with the General Electric Company, Bendix Corporation, and the most recent 24 years in industry with the Exxon Mobil Corporation. In this position, he was a worldwide lead fixed-equipment engineer with his group troubleshooting problems. Since retirement, he has embarked on a career as an author, lecturer, and consultant. He is the author of over 100 technical papers and similar articles. He has written three books on machinery, fixed equipment analysis, and engineering based on his work that have been used for his seminars and consulting assignments in eight countries.

Dr Sofronas graduated from the University of Detroit with a Doctor of Engineering, Pennsylvania State University, a Masters of Engineering, Northrop Institute of Technology, a Bachelors of Science in Mechanical Engineering, and New York State University at Farmingdale, with an Associate of Applied Science in Mechanical Power Technology.

He has been registered as a Professional Engineer in Texas for several decades and had also been elected to the honor society Tau Beta Pi. He received the Society of Manufacturing Engineers, Young Engineer of the Year award in 1980 for his extensive research into the drilling operation for the Bendix Corporation. His doctoral thesis work "The Formation and Control of Drilling Burrs" has been considered a pioneering analytical work into the drilling process.

PREFACE

This book is a compilation of many useful techniques I have learned as a company engineer working in several industries and then after retirement as a self-employed consulting engineer. In it are rules, guides, and examples of what to do and what not to do. There are many personal stories to help illustrate certain points and why the survival techniques suggested are necessary. These personal stories are a way of presenting information in a consolidated way. A story tends to be impressed in one's mind better than rules or guidelines. Everyone will interpret and envision the story in a way most helpful to them.

Colleagues are those who can make the workplace a pleasant environment especially when they have a sense of humor, so some of this is in the book too.

Much of my career in engineering has been involved with determining the causes of failures on machinery and other structures. The term failure denotes something has gone wrong. I like the definition of a failure as an opportunity to better understand it, fix it, and do it right so it won't reoccur.

While the book eventually looks at specific problems solved on machines and equipment using a "niche," pronounced "neesh," it will begin by discussing this niche and the importance of having one. More importantly, it discusses lessons learned throughout a career. They should be most useful for the practicing engineer to know and to help them be successful.

Many case histories have been presented in my previous books [1, 2], and this book adds a few new ones to emphasize helpful techniques and methods.

This is the type book I wish I had available when in industry. Someone asked, "How long did it take you to write this book?" and the answer was 48 years. You see you need the industrial experiences to be able to write a book such as this. The experiences are personal and the cases unique and not repeated from other sources. The readers will now have this information early to help them through their career.

What is a successful engineer can mean different things to different people. Being highly respected and confident, performing useful work, building things, being able to touch things you have built or repaired, solving problems, becoming a manager, enjoying a high salary, and always being employed are definitions of success some use.

For me, success was doing work that I had a passion for and am still doing. I find pleasure in working with others that have the same passion. Leaving adequate time for my family and enjoying them is a huge part of being successful. Advancement and the amenities that came with it were because of this passion. Always wanting to learn something new in engineering is so important. One should never stop learning since that's what keeps us mentally alive.

In this book, I also review my experience with the merits of advanced degrees as many engineers have asked about this. Should they go on for one is usually the difficult decision they have had to make especially if they are married with children. Primarily, their question is if it's worth the effort and some perspective on this is provided.

Should I change jobs, start my own business, or go into consulting are also questions I have been asked and even asked them to myself. In a humble way, I try to provide advice on these important issues.

The intention of this book is to have the reader think of it as a personal mentor and friend always ready to provide help by just opening to the section needed.

ANTHONY SOFRONAS

March 24, 2016
Kingwood, Texas

REFERENCES

1. Sofronas, A., Analytical Troubleshooting of Process Machinery and Pressure Vessels, John Wiley & Sons, 2006.
2. Sofronas, A., Case Histories in Vibration and Metal Fatigue for the Practicing Engineer, John Wiley & Sons, 2012.

ACKNOWLEDGMENTS

First I wish to thank my dear wife Mrs Cruz Velasquez Sofronas, who is a huge part of my success. Most of my advancement would not have been possible without her understanding and guidance. Her review and suggestions on this book were extremely helpful.

Our children Steve and Maria followed us through our many moves and adventures and whom we are very proud of. Both have obtained their college degrees and are having successful lives and careers.

My mother Irene Lampesis Sofronas and my father Steve Sofronas for instilling in me my moral and work ethics as well as for directing me toward my engineering degree.

My sister Carole Sofronas Paquette, whose creativity, writings, and artwork have always inspired me.

Mr Richard S. Gill, my colleague and friend, whose humor and technical abilities made my work enjoyable.

Mr Heinz Bloch, a master of machinery and a friend, for all his unselfish encouragement and for introducing me into the world of consulting and writing articles.

Mr Geoff Kinison, for his friendship, technical abilities, and discussions we have had.

Mr Martin Hapeman, who mentored me with his superb analytical abilities.

Dr Khalil Taraman, my Doctoral Advisor, whose guidance and commitment was instrumental to my achieving the D.Eng.

Dr William Spurgeon, my Industrial Advisor, who directed me through my Doctoral funded dissertation and introduced me to precis style writing.

Dr Paul Paslay for proposing to me an insightful and unique analysis method for my doctoral thesis and discussing simplifying techniques.

To the many superb Engineers, Technicians, Machinists, Operators, Managers, and Friends who have helped me immensely through the years.

To all the legendary engineers whose books I have learned so much from such as Drs Timoshenko, Ker Wilson, Den Hartog, Spotts, Faires, Roark, and many more.

ANTHONY SOFRONAS

1

GETTING AHEAD

1.1 FINDING YOUR NICHE

At some time early in my career, I learned that I needed a niche. A niche will be defined here as something you have a need for, do quite well, is unique to you, and is something you enjoy doing. When it is done well, it can make you a highly valued contributor during your career. There are many type niches; for example, an artist may be known for a particular type of artwork done especially well such as oils, watercolors, or pen sketches of nature scenes, or maybe portraits. An engineer may specialize in analyzing and designing a certain type of antifriction bearing. It's something they do very well and is desired by others.

I've always enjoyed working on and understanding machinery. As a young man, restoring automobiles and diagnosing why things failed was something I liked to do.

I decided to go to a 2-year technical college and learn a trade of machinery rebuilding, welding, and machining. The Associate in Applied Science degree program I enrolled in was unique in that it also contained considerable mathematics and physics courses and how they could be used to design machines. By the way, if you enjoy physics you will also enjoy engineering since it is a sampling of an engineering curriculum. I decided I wanted to design things, so after graduating, I went on to receive my engineering degree. My only concern on making this choice was that I was a "hands-on" type person. I was concerned that I might spend my career behind a desk doing calculations. Nothing was further from the truth. As an example of doing both analytical and field work, Figure 1.1 shows me early in my career taking vibration measurements in the engine room of a new ship during sea trials. I had performed

Survival Techniques for the Practicing Engineer, First Edition. Anthony Sofronas.
© 2016 John Wiley & Sons, Inc. Published 2016 by John Wiley & Sons, Inc.

Figure 1.1   Taking torsiograph readings during sea trials.

the torsional analysis of the engine–gearbox–propeller system and designed it to be free of excessive forces and was now verifying that the calculations were correct. This was a lot of responsibility and pressure for a young man. It was exciting, and everything turned out well. Looking at the photograph now, I realize how dangerous it was hanging over a rotating shaft and taking data during the rocking and rolling of sea trials. It would never be allowed these days.

While the majority of my career was designing, evaluating, and troubleshooting machinery, pressure vessels and structures that wasn't my niche. Many people do a fine job in those areas.

My niche takes a little explaining. As an engineer who enjoys using mathematics to troubleshoot difficult problems, my mathematical skills were not up to the superb abilities of my first industrial mentor. I marveled at the way he solved problems in the most eloquent manner many times using first principles. Now a calculation is said to be from first principles when it starts with established laws of physics and with minimal assumptions or empirical data. For example, my mentor Marty once analyzed the external loads on a complex gear drive system. He started by developing beam-moment equations from the loads and geometry and integrating them to determine displacements. About 20 pages later, he had the solution that was used to upgrade a machine. I still have his report with the beautiful equations neatly and logically presented.

My abilities were nothing of this magnitude, and I felt never to have achieved his mathematical ability. His skills with mathematics were like that of a fine musician making beautiful music. A true virtuoso that couldn't be duplicated.

I mention this because I still needed to and wanted to have this ability, so I developed my niche. I was always good at simplifying tasks. This talent was used to simplify equipment and systems into a form where relatively simple mathematics could be used to solve difficult problems. Sometimes, this also required starting from first principles such as Newton's second law or the energy equations.

In my mind's eye, I would find myself inside the simple model of the machine watching it operate and would model it from there. I might see myself in the equipment hanging onto a pipe as it vibrates or watching a part turn red as it rubs and wears.

Once while explaining this to my son who is a Graphic Designer, I told him that I was having trouble visualizing what was going on inside a vessel I was performing a finite element analysis on. Figure 1.2 was on my desk and that's how he visualized me.

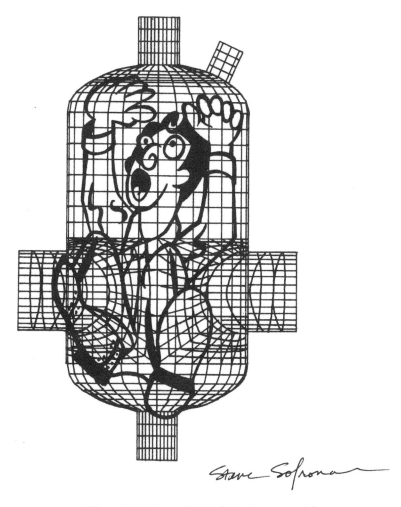

Figure 1.2 Trapped in the finite element model.

The procedure for building a model is fairly straightforward:

1. Visualize, simplify, and sketch the system into the areas that might fail.
2. When there are repetitive elements, reduce them to an equivalent simple system.
3. Make sure the equations include parameters that you can modify.
4. Make sure the failure mode agrees with the data such as a metallurgical analysis.
5. Check the analysis results with other experimental data and be sure it makes sense.

Some of the advantages I've found from being able to simplify systems and build analytical models are as follows:

- A problem can be reduced to a very simple form that is easily explained to others with sketches instead of complex mathematics.
- Modifications can be tested on the model instead of on the actual machine. There is no possibility of a failure if the modification is erroneous. You determine your error on the computer not on the actual system. No one has to know your modification was ineffective and you can easily change it. For example, if you make a reinforcement thicker and the stress is still too high, you change it.
- You can verify your analytical model by using data from other similar failures, tests, or failures found in the literature.
- You can use the analytical model to determine the failure loads and stresses and equipment life. This is like having had a similar failure occur and recording the data.
- You can use all the science and physics you have available as an engineer to develop the model.
- It's extremely exciting to have the actual equipment function like the model anticipated. I found this true with vibration torsional modeling and then having to go do the testing in the field to verify the design modifications I had made.
- As the model developer, you usually have information others are lacking and therefore can provide information to help solve problems.

I've always felt that analytical modeling is the closest I can get to building a time machine. With a good and accurate model built and while sitting at a computer, you can travel back in time to see how a defect could have started to form and then go into the future to see how long it will take to fail. This is truly exciting and amazing especially when it's verified with historical and actual equipment life data. Unfortunately, I haven't figured out how to do this with the stock market. I'm sure someone has, but they are not going to write a book about it and share their secret and neither would I.

Your niche is quite a personal thing. After retiring and writing books and articles on many of the cases I had analyzed I realized not many knew how I had solved the problems. In a working environment when failed equipment is costing your company production losses, all that is required is that you solve the problem. How you solve it is not as important as getting the equipment back in service and explaining what you have done to prevent it from failing again. That's another advantage of a simple model. It allows you to understand what happened, what should be done and to explain this in a straightforward manner to the decision makers.

I'm sure all of you have talents and niches of your own. You should consider developing them further since this is what will help make you unique in engineering.

Sometimes, even if you don't attain the expectations of yourself you were looking for, you find out how to adapt and many times the results are better.

1.2 TWENTY RULES TO REMEMBER

Out of all the work I've published, these 20 rules seem to have gotten the largest positive response from readers and seminar attendees. For that reason, they are stated here again and will be elaborated on in some of the later sections. A colleague provided me with quite a compliment when he commented that they should be framed and hung in every practicing engineer's office.

While rules can't replace common sense or a logical and a methodical approach, they can help avoid embarrassing situations. Here are 20 rules that have been helpful in troubleshooting failures that every engineer or technician will eventually have to do.

The rules have been developed for practicing engineers in the refining industry but should be useful to most engineers and prospective engineers.

Rule 1: Never Assume Anything Making a statement like "The new bearings are in the warehouse and will be there if these fail" is an assumption. They may not be there, may be corroded, may be damaged, or may be the wrong size. The only way you can be sure is to go out and see for yourself.

Rule 2: Follow the Data The shaft failed due to a bending failure, because the bearing failed, because the oil system failed, because the maintenance schedule was extended, is following the data. A string of evidence much in like solving a crime is necessary in problem solving. When trying to solve problems, the person with the data will be the one who can solve the problem. Without data all one has is experience, speculation, or guessing, all which can result in the wrong answer if it doesn't support the data.

Rule 3: Don't Jump to a Cause Most of us want to come up with the most likely cause immediately. It is usually based on our past experience, which might not be valid for this failure. Contain yourself and don't do this and compile data first. This occurs most often when there is a large meeting and everyone is trying to provide input. Be careful when someone of importance or someone who should know does this. Without data, it can short circuit the problem-solving or troubleshooting effort, and focus on only one cause when there may be many interactions.

Rule 4: Calculation Is Better Than Speculation A simple analysis is worth more than someone who tries to base the cause on past experiences. Many an argument in meetings has been solved by going up to the board and performing a simple calculation. It's hard to argue with this type data. Remember engineering is performed using numbers and anything else is just an opinion.

Rule 5: Get Input from Others but Realize They May Be Wrong Most want to be helpful and provide input as to the cause; however, it may not be credible. When interviewing operators, machinists, and others, there are sometimes personal factors that enter into what people say about the cause. This is especially so when one person doesn't get along with another. You need to be aware of these conflicts when collecting data.

Rule 6: When You Have Conclusive Data, Adhere to Your Principles Safety issues are a good example. Your position may not be readily accepted by others because of budget, contract, or time constraints. Before taking a stand, it is important to have other senior technical people agree with you because it could affect your career.

Whenever there are critical decisions to be made, that's the time to be part of a team or form a team to make these type decisions. You don't want yours to be the only name on a document. Engineering decisions are by necessity based on assumptions as all calculations have assumptions built into them.

Rule 7: Management Doesn't Want to Hear Bad News Don't just discuss the failure and the problems it can cause. Present good options that can also be used at other plant locations to avoid similar failures. You will not be popular if you don't have solid methods to correct the problem. You may not need to select which is the preferred option, but you should have the advantages and disadvantages of each. The meeting will be a success if one is chosen or if a next step is outlined.

Rule 8: Management Doesn't Like Wish Lists Only present what is needed not what you would like to have. Adhering to company standards or national codes is usually a wise approach. There are meetings where someone tries to tighten up specifications due to their experiences. The specifications were tighter than recognized national standards or codes and increased the project cost significantly. This didn't go well for the engineer and he was not asked to be part of future projects, which was damaging to his career.

Rule 9: Management Doesn't Like Confusing Data Keep technical jargon to a minimum and present the information as clear as possible with illustrations, photographs, models, and examples. Keep the presentations short and concise. All too often we are proud of the analytical analysis we have done and think everyone else will be too. Most of the time, management just wants the results and what to do next. Details of the analysis are best left to the final report or a trade journal.

Rule 10: Management Doesn't Like Expensive Solutions Only present one or two cost-effective solutions with options, costs, and timing. That is our responsibility as engineers. Present the options best for solving the problems even if the next step is more testing to gather additional data.

Rule 11: Admit When You're Wrong and Obtain Additional Data This is most difficult to do but when other data contradicts yours, it must be done or you will look foolish. In this book, it is mentioned that it is a good idea to have the metallurgical results of a failure available before you present your mathematical analysis. Early in my career I had done this in reverse once, and the failure mode was different than what the Materials Laboratory later determined. The laboratory results were correct and I had to correct my report. It was difficult and embarrassing to do, but it had to be done.

Rule 12: Understand What Results You Are Looking For The analysis was to determine why the rotor cracked, not to redesign the machine. Too often we get so involved in the analysis and forget to just solve the problem. This is especially true for very complex analysis.

Rule 13: Look for the Simplest Explanation First A mechanical engineer might see that a new drive belt was installed too tight and broke the shaft. Computer troubleshooters look to see if the devices are plugged in. Automotive experts make sure there is fuel in the tank. You can then proceed to the next simplest and least costly fix.

Rule 14: Look for Least Costly and Easiest Solution You need to understand what caused the failure first. For example, if a drive belt was too tight, train the machinists the correct tightening procedure. Put a placard on the equipment with the procedure and a caution.

Rule 15: Analytical Results, a Test, or Metallurgical Results Should Agree When the metallurgical analysis says it was a fatigue failure and your analysis says it was a sudden impact, someone is in error. They should both indicate the same failure mode. This was discussed in Rule 11 and shows what can happen if you don't have them agree.

Rule 16: Trust Your Intuition When you feel something is wrong but can't prove it, it's time to do an analysis and get additional data. Your intuition is that little voice in your head that says that this doesn't seem right. All the wiring in your brain store data and observations you have long forgotten, but they are still locked away. So when a shaft looks too small in diameter or a motor looks too small to do the job, you have unlocked a past experience or something you have read.

Rule 17: Utilize Your Trusted Colleagues to Confirm Your Approach Talking with engineering and field colleagues has been the most useful method for finding the true cause of a problem. I usually go out of my way to watch how a job is done or an analysis is performed. After performing an analysis or a design, have someone review the critical ones.

Rule 18: Similar Failures Have Usually Happened Before It is your job to survey your company and the literature for the cause of these type failures and see if it is useful data for troubleshooting this failure. Most pieces of equipment are fairly generic and experience similar type failures. A plant might have several hundred centrifugal pumps. Somewhere in the plant someone has made a repair to prevent a failure. For example, hot alignments on certain type pumps. It pays to be aware of what others have done.

Rule 19: Always Have Others Involved When Analyzing High-Profile Failures When safety, legal, or major production issues are involved, it's unwise to make critical decisions on your own. This is the time for a team approach so that nothing is missed and you have others involved to develop and implement the final solution.

Rule 20: Someone Usually Knows the Failure Cause It has been my experience from interviewing engineers, operators, machinists, and technicians that several usually knew the true cause of a failure. A good interviewing procedure is therefore an important part of troubleshooting. For those that know the solution, give them the credit they deserve.

1.3   CALCULATED RISK VERSUS REWARD

As engineers we like to limit our risks. As a general aviation pilot and mechanical engineer, this has served me well over the years. I didn't do things that were too risky and always had a couple of alternative plans in case something went wrong. For example, when flying cross-country, I always had alternate landing sites in the event that the weather deteriorated. In design, my request for design modifications was always supported with adequate calculations. When someone has done a reasonable analysis, their arguments usually carry more weight than those who are speculating on the cause with no supporting data.

There can be problems with this approach. There is always risk involved in every engineering decision and you cannot progress far in your career if you are unwilling to take some risk.

Consider a large steam turbine vibrating slightly above normal levels with blade fouling thought to be the problem. Management wants to know if it can be run 1 week until a planned outage can be scheduled as thousands of dollars in profit a day are at stake. Your career will not be enhanced if you say it has to be shut down immediately, with no supporting data. Likewise, this is not the time to try your first attempt at online washing of a large steam turbine while it is in operation. This is risky business if you have no experience and no operating guidelines for the procedure. However, this would be a good time to monitor the vibration level, talk with the manufacturer and others with similar machines and then determine the risk in just monitoring the vibration levels. Defining at what vibration level it will have to be shut down will still require some risk, but now others are involved. The reward for doing an online washing yourself and being successful will make you a hero and elevate your status in the company. The risk is wrecking a million dollar machine because of your lack of knowledge. You would never be able to recover from this judgment call in this company and would probably limit your career growth, meaning you would not be trusted with decisions. I don't know about you but to me the reward is not worth the risk. I'd rather be around with the company to solve the next problem.

Obviously, there is much more to this in the judgment-making process, but this illustrates the need for some calculated risk.

1.4 ADVANCEMENT

Salary increases or raises are something we all expect when we do good work. Early in our career, they tend to occur fairly regularly with your supervisor coming to your desk with a slip of paper or it just shows up in your pay check. They are nice and show that your work is noticed and appreciated. The frequency of the raise is built into your department's budget. How much goes to each person in the department, if any, is something the department manager has to figure out. I have had to do this and it is a difficult task that I took very seriously. When someone didn't get a salary increase periodically, it was never a surprise to them because the reason was always in their performance review, which we had gone over. What they had to do to improve was also in the review.

Promotions are different and require much more consideration. When you are promoted, your responsibilities change. A company has a limited number of these positions, and there is usually considerable competition for them. The darker side of corporate politics starts to appear such as favoritism and resentment by others. For higher level promotions, it is sort of like running for a public office and you will need people in your corner.

With these promotions, you should receive a substantial salary increase and other benefits. Along with that will be new responsibilities and the requirement that you develop new talents, more travel, and longer hours with an increased work load. You cannot expect to be promoted and not do more. However, the satisfaction you receive is usually well worth it.

The best way you can understand the requirements of the new position is to look at someone who has that title in your company and realize you would like to do the job better. What would you do, and what are your goals for yourself in that role?

We use to always be amused when a new Vice President (VP) of Engineering was brought into a company. When you are at a prominent position and come into a new area, it seems to be imperative to make yourself immediately known in some way. In this company, the tradition was to paint the offices a different color, say from yellow to pale green. The next VP would paint it from pale green to yellow. It was always fun to watch and occurred several times.

The changes in my titles weren't quite as prestigious but still required that I do something different. One position had me directing a troubleshooting department. The first thing done was to analyze all of the technical and analytical capabilities of the new group and make up a one sheet list on what each of them was expected to be proficient in. This would then be used when visiting our customers at the company sites.

When opportunities were available, the talent in the department was ready to assist the plant. Making sure the individuals went to the correct seminars and also stayed current with the complex software used were also included in the goals for the group.

Once my long-term goal was to be Chief Engineer for a large company. I didn't care for being in management and wanted to stay in the technical arena. This seemed to be the top rung of the technical ladder in the company.

I never got there and after seeing what it required was pleased that I hadn't. I didn't really have the long-term vision for that position nor the communicative skills necessary to work with top management. This position required one to be aware of all activities around the world that were going on in your area. You were responsible for things that went wrong even though it wasn't your fault. You were responsible for the higher level promotions, long-range planning, budgets, equipment improvement programs short- and long-term, personnel issues, and much more. You may have noticed that what is missing are the things I was best at and enjoyed the most, analytical modeling and troubleshooting. The Chief Engineer should assign others to do that type work.

I probably could have gone to a smaller company and been a successful Chief Engineer since I could have performed technical work, but I was quite happy in the company I worked for. There were plenty of opportunities to stay technically challenged. The title, benefits, and prestige would have been wonderful, but the work not as fulfilling.

So with title changes come increased responsibilities and different type work that you may or may not be comfortable with. You will need to make that decision when the time comes.

1.5 LEARN FROM OBSERVING FAILURES

The term failure is not a politically correct term, and lost opportunity, disappointing outcome, or errors in judgment might be more palatable; however, I prefer to use failure. A failure to me means that the outcome of an endeavor wasn't what you intended it to be but represents an opportunity to correct it and do it better.

I once heard and now strongly believe that the only failures in life are those that you have not learned anything from and repeat them. We all have had failures both in our professions and in our personal lives. Some people can be crippled by them and never recover. Others take them as a tough learning experience and tell themselves that they will never let that happen to them again.

Early in my career, I had gone from receiving my doctorate degree in engineering right into becoming Manager of Advanced Engineering. While I didn't know it or was too naive when I accepted the position, this new job eventually required me to reduce the size of the department, meaning letting people go. This was something which I just couldn't do. I left the company after 2 years and went back to doing technical work, which I have never regretted. This could have been considered a failure, but I never did for several reasons. First I learned a lot about myself and what I could and couldn't do and was proud of my decision. I had spent my career developing my technical expertise but was not able to use it in this position. However, I did learn about forming a test laboratory, developing a yearly budget, funding new projects, developing work plans, developing performance reviews, working with high-level customers, and mentoring personnel. This was all extremely useful in my future technical career when I directed a worldwide problem-solving group. So I took something that could have destroyed confidence in myself and used it as a very constructive learning tool. As Winston Churchill once said, "Success is the ability to go from one failure to another with no loss of enthusiasm."

There were not many technical failures since critical decisions were never based on my opinion alone. Input from others on the final design or final modification was always requested. Work was always supported with calculations and historical data and not only with experience. As you will see in later chapters, the ability to develop analytical models to better understand the operation of the equipment was immensely helpful in keeping decisions quantifiable.

Of course, sometimes the analytical models might have been too simplistic or there may have been some important data left out, but there was always the opportunity to learn and improve.

My first job in engineering was working for a company that produced the drive wheels for those big 200-ton off-road vehicles used for mining. They were immense, bigger than some houses and the wheels were huge. I had been working for the company for about 3 years when the engineering manager called me into his office and said he had a job for me. It seems that there was no oil level indicator for each of the drive wheels and the equipment owner wanted them for maintenance. They wanted the driver to be able to see how much fluid was in the transmission before each work shift. The job he said was to design them for field installation on the various vehicles that had already been shipped.

This was not a simple job since there were several type vehicles. Also there were several variations of each and all had different sump arrangements. It took several weeks of reviewing all of the design drawings at the factory and having various welders fit the design to a wheel in the shop. Fellow engineers then came down to see if they could follow the field installation instructions that had been written.

When I felt comfortable with the design, I went into my supervisor's office, put the design and plans on his desk, and told him it was done. To my surprise, he then handed me a plane ticket to an open-pit copper mine somewhere in the western part of the United States and said to go show them how to install it. That was not in my plans.

Explaining to the mine manager why I was there was interesting. He looked over the plans and listened to me with a smile that was disturbing. He then called into the office this 6-ft, 300-pound welder named "Tiny" to install it on Unit 276. When we got to the vehicle, I understood the smile. The vehicle was parked in the normal 4 ft of mud. The top of the level indicator cap would be at 3 ft, meaning 1 ft below the mud. Quite difficult to check the oil level with the arrangement designed. The only thing more embarrassing was the comment received from Tiny. "Do you want me to install it now or would you like to think about it for a while?"

I bring this example up because it shows the importance of actually seeing the equipment you're designing and talking to those who will be using it. You should see the equipment under its worst-operating conditions. Like all embarrassing situations, you realize what you should have done 1 µs after you see what you did wrong. The important thing is to learn from it and not to let it happen again.

You can expect some failures during your career, but they should not all occur early in your career, be consecutive or major.

1.6   KEEP GOOD RECORDS OF WHAT YOU HAVE DONE

Here we will review why it's so important to document failures you have seen.

I'm not a materials person and send most of the metallurgical and nonmetallic work to laboratories for analysis when performing consulting jobs; however, there are preliminary observations engineers can make.

Here are some after the fact analysis meaning there was a good explanation for the cause of the failure. All that is needed for the analysis is a little knowledge and some magnification.

Figure 1.3 represents part of a disk dryer assembly in which two disks were held together with plug welds. A plug weld is a weld from one side only when the other side is inaccessible and in this way leaves a joint with a gap that is susceptible to

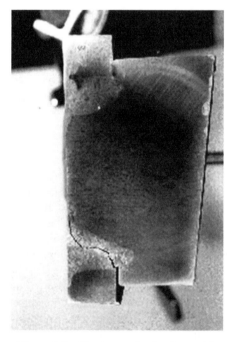

Figure 1.3   Crack growth of a plug weld.

crack growth if excessively stressed. The outline of the weld is evident and has been ground smooth on the left side. The problem with a plug weld is that there is a stress riser where there is no weld material. A fatigue crack can be observed starting from there.

Figure 1.4 is the failure of a titanium connecting rod for a racing car. The fatigue failure is due to a faulty design. Just looking at the failed pieces shows the small piston pin portion had an oil hole in it. It was too thin a design and caused a fatigue crack to grow from the oil hole.

Figure 1.5 represents a large coil spring failure. This was one of several that were failing on a large vibratory conveyor. The failure was near the first small coil where it was bolted to the structure.

It would be quite logical to think this spring might have been overloaded and cracks started on the corroded surface pits. Additional information came from a hardness check of the springs, which indicated this batch of springs had been incorrectly heat treated due to a new supplier being used.

Figure 1.6 is a bolt that has undergone an impacted type bending load. This was one of four bolts in a large mixer that was struck by large chunks of product. The product fell off of the vertical baffles and the blades impacted it much as a baseball hit with a bat.

Figure 1.7 is a stop-drill hole in a stressed plastic piece but could just as easily have been metal. Stop-drill holes are small holes drilled at the end of a growing crack as a temporary repair until a permanent one can be made. The theory is that the drilled hole has a stress riser much less than the radius at the tip of a crack and therefore should halt the growing crack. Figure 1.7 indicates why this is only a temporary repair. At some point, this part was highly stressed again and the hole itself acted as a stress riser causing a secondary crack to start from it.

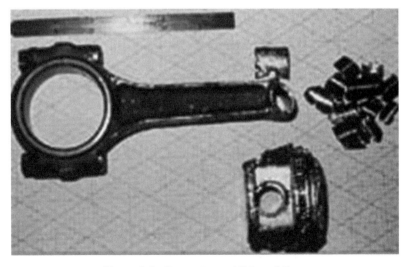

Figure 1.4 Connecting rod fatigue failure.

Figure 1.5   Spring failure.

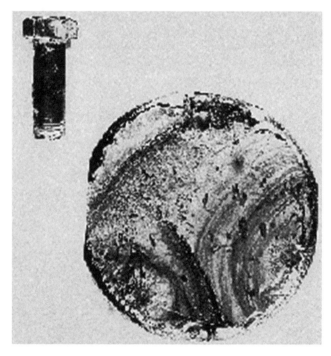

Figure 1.6   Impacted bolt.

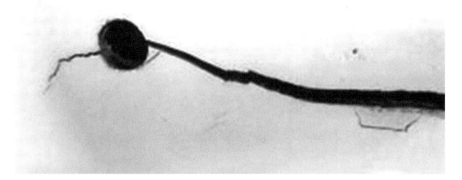

Figure 1.7   Secondary fatigue crack growth.

Figure 1.8   Thermal cracking tube ID.

Figure 1.8 represents thermal surface cracking on the inside diameter of a high-temperature furnace tube. The cracks were 1/8 in. deep in a 1/2 in. wall thickness. The outside diameter indicated no cracking. These type failures require a detailed examination by a metallurgical laboratory experienced with high-temperature materials. However, seeing cracks such as this should alert the engineer that this is a deviation from the norm.

The key learning here is to document all the failures you have observed because sooner or later during your career you will most likely see them again. Knowing what the cause was can provide valuable troubleshooting information.

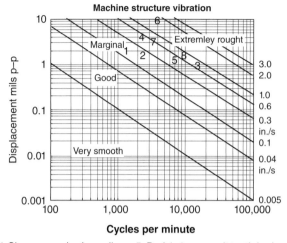

Machine structure vibration

1. Slurry pump bad coupling
2. Bad agitator
3. Bad screw compressor
4. Steam turbine resonance
5. Bad 1:1 gear unit teeth broken
6. Imbalance centrifuge
7. Vertical multi-stage pump – bushings
8. Small trubine missing blade

Figure 1.9 Vibration test data [1].

When an analytical calculation is performed or data is obtained, it is recommended to put it in a report, plot the data, or put it in a table of some form. In this way, it will be readily available when a similar failure occurs, which they usually do.

Figure 1.9 illustrates this with some data from vibration testing on equipment. When additional data is obtained, it too can be superimposed on the graph. Note that some details on the cause of the failure are also recorded for future reference.

When you perform an analysis and later find out what the problem was and if you were correct, one approach is to tabulate the data. Table 1.1 illustrates this for bolted joint calculations.

TABLE 1.1 Bolt Loading Calculation History

Application	Bolt Diameter-Length (in.)	Torque (ft-lb)	Relaxation/Stretch	Preload/Alternating	History
Vibrating conveyor weight	3/4–10	200	0.002/0.012	14.2	Ok
Agitator paddle	1–3	150	0.001/0.001	1.1	Failed bending, found loose
Agitator paddle	1–3	500	0.001/0.004	4.3	Ok
Cutter blade	3/4–2	275	0.003/0.003	4.9	Bolts loose and broke
Cutter blade	3/4–2	450	0.003/0.005	8.0	Ok
Main bearing cap	5/8–2	5	Loose	0	Failed in fatigue

The table is much larger and as shown is only a sample. From this chart, important bolt calculations and the changes made have been recorded, and it is evident what made them fail and how the modification worked out. This is valuable historical data if a similar failure occurs.

1.7   FLEXIBILITY IN YOUR CAREER

An acquaintance that was a professor of law said that some of the best students he had in his classes had engineering degrees. This wasn't surprising since logical thinking, research, problem solving, and reasoning are some of the things engineers do routinely and most do well. One question to him was why they had chosen to go into the law profession. He said there were many reasons one being that it would be an excellent second degree as it would open even more career opportunities in engineering such as corporate law or as an expert technical witness.

This is not the type of flexibility I mean. The majority of engineers in industry are doing something other than what they received their degree in. Electrical engineers are working as mechanical engineers, computer specialists are process or systems engineers, and as for myself I've worked in multiple areas. One tends to migrate to areas where the job opportunities are. Someone who is an environmental or marine engineer has to know about thermal effects (thermodynamics) and how to design, modify, install, or use instrumentation and computational systems. An engineer developing prosthetics has to understand dynamics, stress analysis, and robotics to name just a few areas.

At one point in my career, I looked at the CEOs of several companies and realized they all started out with engineering degrees. They may never have practiced engineering for more than a few years until their leadership talents were realized, but the strong technical problem-solving and decision-making background was there.

I have worked on a preliminary design for a Martian soil sampler, which was a research laboratory proposal to NASA, locomotive and ship engine designs, aircraft braking systems, racing car frames and engines, refinery equipment design, and manufacturing product testing equipment. I have been troubleshooting various types of machines and structures as well as spending a couple of years as Manager of Advanced Engineering for an automotive products manufacturer, taught at a university and presented seminars on most continents. After retiring, I started my own engineering consulting business not because I had to but because I enjoyed engineering and the people so much that I wanted to continue at it. This just shows the wealth of opportunities and flexibility available with an engineering degree.

1.8   YOU'RE KNOWN FOR YOUR WORK

What you publish distinguishes you from others. A report that is concise and fact filled with new information and a minimum amount of fluff is always welcome. Fluff will be defined here as information that is obvious or readily available and after you read it, you feel as if you have wasted your time. Precis type writing and newspaper headlines have no fluff as every word means something. More about that in Chapter 4.

When someone writes a technical paper and in it states the load is approximately 112.213 pounds force, this is troubling and makes you doubt the person's capabilities. Why has "approximately" been taken to three decimal places? Material properties alone can have variations of $\pm 10\%$ so the decimals have little meaning.

Figure 1.10 is fictitious, but similar ones have been seen in magazine articles and are especially troublesome.

The problem with this graph is that if you calculate the regression coefficient R^2 it is less than 0.2, which means there is no correlation between time and load based on this data. Therefore, the trend-line makes no sense. While in some fields low R^2 occur such as when evaluating people, this is not the case in engineering. So what is the author of this paper trying to do with this graph, certainly not use it to determine a relationship between time and load for this process. The data might be useful without the trend-line if the author is trying to show there is no relationship or the recording instrumentation was in error but unfortunately this is not usually the case.

Here is one more case slightly exaggerated but seen in articles. Figure 1.11 is used to show a "sudden" stress increase over a period of time as determined by strain gauge stress tests.

The problem, of course, is the truncated scale. If the full stress range was shown, a 400 lb/in.^{2} change would hardly be noticed as a bump. It is quite unlikely that any

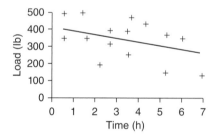

Figure 1.10   Scatter plot of load versus time.

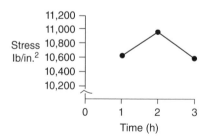

Figure 1.11   Stress–time.

strain gauge testing over this time span could be this accurate. Even if it was, what difference would this small change in stress make?

Many experienced engineers might stop reading the article as soon as data like this appeared and discount the rest of the article. It's important to understand how your readers will interpret your results.

1.9   ETHICAL BEHAVIOR IN ENGINEERING

Each professional society has its own code of ethics that engineers are expected to follow. They are similar and coincide with what most employers also subscribe to. For example:

- Hold safety paramount on any endeavor.
- Be aware of conflict of interests.
- Be aware of confidential and proprietary information.

There will be times in your career when you are aware of safety issues but don't know how to present your concern. This was certainly the case with some engineers [2]. My experience with this is only from reading about past disaster reports that have been published by government agencies.

A basic ethical dilemma occurs to an engineer when there is a possible safety risk that the engineer is aware of or because the engineer's instructions have not been followed. For a consultant, if they do not report this to the authorities, they may have their license revoked. For a practicing engineer, there may be severe consequences to their career especially if there is a major safety concern that eventually causes a loss of life.

Sometimes, going around the internal management of the company to report such deviations is called whistle-blowing. From the reports read, this never went very well for the whistle-blower and it is not recommended unless you are looking for other places of employment.

It's our obligation as engineers to report such safety concerns. Personally, I wouldn't sleep well if I didn't. I would never ask anyone to do something or operate something I wouldn't be willing to do myself.

An approach to solve such a dilemma would be to provide calculations to illustrate what could happen. An example of this is shown in Section 10.10. Here concern was with a safety issue on testing a vessel. This approach was used to discuss how the team could perform the test in a safe manner. It was agreed on and the test was performed without incident. Thus, no so-called whistle-blower situation occurred and the concern was addressed and corrected most amiably.

What would have been done if this didn't convince the project team? Working your way higher up on the management ladder until someone listened was a possibility. What would have been done if no one listened? Document your findings as this would have assigned responsibility to those who would not correct the problem. Most will do something when safety issues are their responsibility. This could have some repercussions as this is a fairly severe step to take. Fortunately, things never usually got to this stage during my career.

Another case was when a colleague Brian reported that in an industry committee meeting he was a member of, someone had reported a safety concern on cyclic pressure vessels used in industry. Since these were high cycle pressure vessels subject to historical catastrophic fatigue failures, he said we should research the subject as it was both a materials and mechanical engineering problem. We decided the best approach to document our findings was to present it to a worldwide forum [3]. After consulting with our company's legal department, we did this with their approval. We had such vessels in our plants that turned out to be nonissues. Others in the industry weren't as lucky. My colleague was to be commended on recognizing the safety issue and wanting to share it worldwide. Our company was to be commended on allowing us to do this.

As professionals it is the duty of engineers to perform our work with the highest standards and integrity utilizing the appropriate tools. We should recognize the value of inputs from our colleagues to insure our work upholds the prestige of our profession.

1.10   HUMOR IN THE WORKPLACE

I've always thought misery would be having to spend time with noninteresting people who had no sense of humor since it would make for a very long workday. Luckily, this never seemed to happen during my career. When I retired, I wrote a book entitled "Family And Friends In The Oil Patch," which was self-published. It told stories of many of the people in my life and their humor. A copy was presented to each person at my retirement party since they were all in the book. I even learned a little illustrating for the book since some of the stories could be better visualized with a cartoon. You see during my career I enjoyed hearing stories people told about their experiences and would write some of them down. Many of these were heard in the machinists break room or with operators in a control room while gathering data for a failure analysis.

An example might be Arlon. Someone at work might ask me what I was waiting for and the answer might be the evening incident report. Asking Arlon what he was waiting for and he'd say, "5 o'clock." I've worked for several companies all over the country and each area has its own specific humor as does each occupation. It was fun to experience them all.

Once I was called by a colleague who resided somewhere in the mid-part of Canada, where it was -20 °F outside. He was responsible for a gas transmission site. The site had about 20 gas-engine compressors in a long compressor house with doors on each end and compressors on each side of the corridor. It looked something like Figure 1.12.

In the middle of the telephone conversation he said, "Hold on a minute." From the background noise it sounded panicky with many people yelling and alarms going off. When he got back on the line he said, "Sorry about that we just had a moose running around in the compressor house and we had to get him out." Seems like someone

Figure 1.12   Typical compressor house.

had left one of the doors open and the poor old moose just wanted to get warm. You remember things like that.

Some of these stories came in handy when presenting seminars on a subject and they were relevant. They lightened up the presentation since so many of the attendees had witnessed something or someone similar.

We tend to remember people who didn't take themselves too seriously and could laugh a little at themselves and the things they had done. Each of you has had similar experiences and it would be good for you to write them down, least you forget them.

1.11   SELF-PRESERVATION WHEN DOCUMENTING YOUR ANALYSIS

This subject is so important it's in the beginning of this book. It's a caution every engineer who performs failure analysis work should be aware of.

When you are analyzing catastrophic type events such as pressure vessel explosions or machinery impeller containment issues as discussed in Chapter 10, certain precautions are almost mandatory.

First, engineering is not an exact science, and for any analysis, assumptions have to be made. The more assumptions the less exact the results. Even material properties have a statistical range of possible values. Therefore, any analysis you do can always be contested in a court of law by an engineer brought in by the opposing council.

For these reasons, when many assumptions are made in an analysis such as the mode of failure or size of the debris generated, I usually will not publish the analysis with the failure. Only the results of the analytical model are used to justify the suggestions and recommendations made as shown in Chapter 10.

Friends have asked me to analyze modifications made to various engine designs. This is usually done, but the calculations won't be provided to them. After performing the analysis and if the design appears faulty, the conversation with them might be something like, "It appears to be a little weak in the rod pin area and you should have the manufacturer redesign it" or "Make sure you have a detailed periodic inspection procedure in place for the rotor disk so there are no cracks." This would then be discussed with them verbally with some details provided.

You see that as an engineer you may be the only one who has done any analysis on your friend's design. It's just a small part of the design, but you may be thought responsible for more. Even though you didn't evaluate the rest of the design if a safety incident occurs, you may be issued a subpoena and required to be in court as the sole engineer involved. This is also true for the Professional Engineering license. Don't stamp documents that you haven't done the complete design for as you may not know all the details.

When you are doing work for your company, you usually know the people involved and their capabilities. You can go out to the failure scene, interview people, and look at the parts and the metallurgical results. As a consultant, you probably have incomplete or incorrect data so any analysis you do may also be in error. So when the opposing council says "Are you sure you had all the correct data for your analysis?" your answer will have to be "No". And there goes your council's defense and the reason for having you there.

We are usually proud of the work we have done and want to present it. Unless we are using recognized national standards or code calculation software, we need to be aware of the difficulties presented here.

For these reasons, a formal problem-solving method on catastrophic type failures is recommended. Here, there is a team moderator and team members and you contribute to the team and follow a prescribed failure analysis methodology. You are one of several presenting the final recommendations. Here you can show the team your calculations and why you made the recommendations. Now this is a much more self-preserving way to present your analysis and have it implemented with little risk to you.

1.12 DON'T BE OVERWHELMED

When we are asked to solve a difficult problem, it can be overwhelming. Observing a major failure of a complex machine or structure and realizing the magnitude of the problem can be intimidating. Looking into a gear box as large as a truck with all the teeth sheared off or a hole in the side of a compressor with pieces scattered all over the site and the task to determine what happened can appear overwhelming.

That's the way I usually feel before I have any data. In engineering, data is everything and without data we are just guessing.

That's what's so wonderful about engineering; we can obtain the data or generate the data. We obtain it with research, interviews, observations, and collections. We generate data with measurements along with metallurgical and analytical analysis. A hopeless pile of rubble soon becomes a trail of evidence for use in determining the cause or causes.

As long as our approach is organized, logical, and well planned, the foggy nature of a failure slowly begins to clear and the solution becomes evident.

Don't be forced into producing a cause for the failure too early as you will probably be wrong. It's usually better for your career being criticized for taking too long and being correct than arriving at a solution quickly and being wrong. Take time to collect and organize the data as suggested in Chapter 5.

We have all heard the initial national news reports when a catastrophe happens. The knowledgeable people wait until there is some data to report on and those who don't have any idea speculate on the cause. When the true cause or causes are found, those who speculated and were wrong look pretty foolish.

1.13 PROVIDING GUIDANCE TO OTHERS

Most of us will have to provide directions to others on performing a job. The task might be designing something, performing a calculation, or what to do when selected to be on a major project.

Some of us want the person to do the job as we would do it. The problem with this approach is that it might discourage the person from doing it their way. By using their own initiative and not following a "cookbook" approach, a better result may be realized along with the satisfaction of having come up with the solution on their own.

To allow this to happen, I sometimes take the following approach. I will provide general guidelines, such as previous work done or a good book or article along with the necessary cautions. They then proceed as they see most reasonable unless they ask for assistance. I check on their progress periodically and if their approach seems like it will fail I either let them fail, if it will not damage their career or make further suggestions. Failure is a harsh but effective teacher.

Unfortunately, there are those who will not follow suggestions and things may not work out well for them.

For example, an engineering consultant I knew was about to perform a complex three-dimensional finite element analysis on equipment my company was about to purchase. Before starting the analysis, I suggested to the engineer that he should go and see how the equipment operated and do a simple analysis before building the model and contact me if he had concerns. He didn't do this and his model was completely in error since he didn't understand the machines operation. Since he was not working for me, I couldn't periodically review his work nor did I think I had to. His analysis was never used and he lost credibility with me and my company.

1.14   THE TECHNICAL AND MANAGERIAL LADDER TO ADVANCEMENT

It has been said that 70% of engineers consider being managers during their career especially those 40 years or younger. After that age many would just like to be on major projects that interest them [4]. That's why this section is added to the book.

During my career in the three large companies I worked for, there was what was called the technical and managerial ladders of advancement. It was said that one would start as an engineer or technical person and then with enough experience could pursue either. It sounds good but the fact is that some of us are just better suited for one or the other.

I've spent most of my career in the mechanical engineering technical area because that's what I found I do best and enjoy the most. The work/life balance has been just right for me. I was manager of advanced engineering for a few years when I was 40 years old after receiving my doctorate degree. I feel the position was available to me because advanced technical work was being done in the new organization. The Doctor title would indicate to our customers the highly technical nature of the team that would be developing their products. I didn't question the reasoning. Many times we take the managerial route because that's where we think the prestige, promotions, bonuses, and salary increases are. Many companies expect promising engineers to take these opportunities when they are presented to them.

This is an important decision point in one's career and it should be analyzed very carefully as the opportunity for such promotions don't occur often. Only about one in fifteen technical folks becomes manager for most of their career. Once you make the decision to proceed into management, it isn't easy to return to a technical career within the same company. When you have been a manager for many years, you will lose your technical edge as things change quickly in the technical world. Also the bonuses and prestige are hard to leave. During the time I was a technical manager, the finite element method (FEM) and computational fluid dynamics (CFD) which I had used changed considerably. When I went back to the technical ladder from the managerial one, it took me a considerable amount of time to regain that technical edge and yes I did have to change companies.

The term technical manager is used here because that's what most engineers consider, meaning management focused on technical developments. It differentiates from say management of businesses or management of people.

Table 1.2 represents some of the major differences between technical managers and technical engineers that I have experienced.

This table certainly isn't complete but it does include areas I've been most aware of. Some of the areas might overlap but usually they don't. When I say management input is better received by senior management, this isn't because there is a dislike of engineers, it's just that managers have the similar thought processes to each other. For example cost, timing, and probability of success. An engineer would be thinking about how the problem was solved. Studies show [4] that over the age of 50 the notion that high performance results in promotions is less believed. I think the organizations' promotion standards, which can change with time, and high-level positions available

TABLE 1.2   Differences Between Managers and Engineers

Technical Managerial Position	Technical Engineering Position
Potentially higher salary	Can build and develop things
More power and prestige	Make decisions based on own analysis
Higher recognition	Become a well-known expert
Higher level advancement possibilities	Publish your technical work
More perks such as bonuses and travel	See things you have designed
Can make major financial decisions	Can become a consultant
Scope out new products not just a piece. An engineer may work on a piece.	During economic downturns can find other job because of engineering skills
Decisions based on input from others	Can retire, start a business, consult, and present seminars because of skills
Can direct the career growth of others	Salary can be higher than managers for senior level engineers
Directs work through others. Engineers do the work	There is always something new to learn and work on in engineering
Input is better received by senior management	There is more out of the office field work and interaction with the trades

control these promotion opportunities. Having a high-level manager interested in seeing you promoted is also a big plus.

In smaller companies, technical managers may also be doing engineering type analysis and a chief engineer would be an example. In a large company, a chief engineer would probably be doing many of the tasks of a technical manager such as scoping new developments and personal development and would, or should, leave the analysis to the engineering staff. This is what happened to me as a technical manager and it was difficult for me watching the engineers in my group do all of the exciting analysis work. I had technical input but it wasn't as rewarding as doing it myself.

I must admit that the technical manager role certainly had more room for promotions as I saw it done with those in the position before me. My boss who was the director of engineering had my job before me and he wanted to move up to be vice president of engineering. The vice president wanted to retire. The job certainly had the power and prestige with use of the company jet and helicopter, bonuses and possibility of making significant new products, and possibly a new division for the company. All the elements of success were there. It just wasn't a career choice I would enjoy. I had developed my career as an engineer and all of my education and experience were in this area. I felt I could contribute much more as an engineer. Others would have been quite content with that managerial position but you have to be true to yourself. Although I missed the bonuses and company jet, I never regretted my decision.

Engineers need to realize that they should have the experience and develop strong technical skills before they accept a position as a technical manager. They will be more respected by the technical staff knowing their manager understands the technical details discussed. In the short time as manager I had gotten along quite well with my group as I understood the details of what they were doing.

So the question then might be, how does one prepare themselves for a managerial position?

It really depends on the company and its policies, but there are some general areas of importance and many have already been discussed elsewhere in this book.

Learning to communicate well with others is important. Don't make enemies of the wrong people or be known as being annoying or weird. Don't have major technical failures associated with you but be known for making sound and logical decisions. Look around for what you consider successful managers, see how they conduct themselves and try to associate and learn from them. Contemplate what you would do different to be a better manager. Learn to work with and through all types of people and give them credit for what they do.

Joining technical societies, developing and presenting technical papers and being on or better yet heading up committees are ways to develop your technical managerial skills. Making the necessary connections with those in power by letting them know of your presence is always wise.

Having your assignments rotated within the company is a good way to have a better overview needed as a manager. It will also allow you to see and meet other influential people. Staying as the key technical person in a small important area can make you too valuable to promote into management. Actually all of this is valid even if you prefer to stay on the technical side of the ladder as it will make you a better educated contributor.

There can be some luck involved. In my case, the managerial position had just become available, I was available and well known and had done some good work for that organization. I had the title they were looking for and my director knew the organization's vice president and recommended me. I probably wouldn't have taken the job if it wasn't presented to me. I was quite happy with my technical position and saw a successful growth plan. This all changed when I was asked to consider the new managerial position.

REFERENCES

1. Sofronas, A., Case Histories in Vibration and Metal Fatigue for the Practicing Engineer, John Wiley & Sons, 2012.
2. Columbia Accident Investigation Board, Report Volume 1, , National Aeronautics and Space Administration, Government Printing Office; August 2003. 248 pp.
3. Sofronas, A., Fitzgerald, B., Harding, E., The Effects of Manufacturing Tolerances on Pressure Vessels in High Cycle Service, A.S.M.E., PVP Vol. 347, 1997.
4. Allen, J.T., Katz, R. The dual ladder: motivational solution or managerial delusion, R&D Management 16(2), 185–197, 1986.

2

THE POLITICS OF ENGINEERING

Many years ago, when I started in engineering, there was a pamphlet from the American Society of Mechanical Engineers about "The Unwritten Laws of Engineering" [1]. It was about engineering ethics and what an engineer should and shouldn't do. Much of it is as valid today as it was when it was written in 1940. It's a 50-page pamphlet, which I last read 45 years ago. I haven't ever taken any ethics courses or seminars on the subject of how to behave as an engineer. After reviewing the pamphlet again, I realized many of the rules and advice mentioned had been followed and several are presented again in this book. Much of what we should do and how we should behave is just common sense. This is the benefit of a book based on personal experiences and case histories. They withstand the test of time. It is recommended that someone new to engineering or engineering management read this or a similar book early in their career.

2.1 WHAT TO DO

There are many key things one should do in an engineering establishment. One of the most successful things done early in my career was to observe senior technical people and how they behaved. The chief engineer didn't get where he or she was by making risky decisions or taking credit for work others had done. If they had, then this would also be good information not to follow.

Keeping your management apprised of what is going on is always wise. Management doesn't like surprises, embarrassments, or things that will affect their career

Survival Techniques for the Practicing Engineer, First Edition. Anthony Sofronas.
© 2016 John Wiley & Sons, Inc. Published 2016 by John Wiley & Sons, Inc.

growth either. Even if you don't approve of their methods, it's usually better for your career to keep it to yourself. I have memories of many who voiced their opinions to their colleagues, in so-called confidence, about management, and it usually worked out badly for them.

Most of the valuable contributions made during my career have been due to contribution from others. This is true for just about everything we do. Giving people credit for their achievements and pointing them out to others has always been welcomed as it's not usually done. Surprise people and do it as it's the right thing to do.

I've always appreciated people who were dependable meaning those that said they would follow up on something and they did. What is really welcome is when they do more than was expected. Your management will value this if you do this also.

Making decisions is something not everyone can do but which is extremely necessary in industry. Many decisions are not difficult if you have problem-solving and analytical abilities and can verify decisions. There is not much "guessing." This type of engineering decision making is intellectual rather than impulsive and is usually backed up with a well-defined plan that is easy to defend.

2.2   WHAT NOT TO DO

One just has to reverse the "what to do" and things will usually end poorly, so there's no use repeating them.

Every engineer should be aware of territories. Just as many predatory animals have their territories in which they hunt, so do certain technical people. When you work in one area for a long time and foster many alliances with departments, some organizations can feel like you are encroaching on their territory and defend it fiercely. For example, there may be a central engineering group in your company who service certain divisions through contacts. You would want to be aware of this before charging in to solve problems. It would be wise to find a more discrete way to proceed.

Likewise for individuals. When you are working on a company problem, you may want to contact company experts in that area. When you recommend a change without their approval and your decision is erroneous, it could be disastrous to your career as you will receive little support from anyone.

There's a fine line here since you don't want someone taking credit for your work and claiming it as their own. Plagiarism of intellectual work is serious and we should defend our work vigorously as it represents our future growth in engineering. One way to do this is to document your work by writing company technical publications and ask for input from others and acknowledge their contributions. A joint title publication would be considered when someone contributed approximately half of the effort. Outside publications work well too; however, they usually need to be released through the legal department, which can be tedious.

In industry, it's good to be respectful to everyone, but trying to always be "friendly" and please everyone doesn't work. There are people who will take advantage of your good nature. So you must learn when to defend yourself and always be a little suspicious. This is not to say you should be stern or pompous.

Previous managers have told me that I could be a "prickly pear" at times and at one time a plant manager told my manager that I was a "polite son of a gun" only he didn't say son of a gun. I was pushing his organization to implement a modification because of safety issues and the solution was going to be expensive. All the calculations, backup data, and support to prove my point was done, but he said he didn't want to spend the money. I told him that was fine, but my recommendations would be documented and the decision was up to him. Since he didn't want the safety issue in his arena, he agreed with much reluctance. You have to be fairly high up the technical ladder so that what you say holds some authority. For a new engineer taking that approach without adequate senior support, the results could have been quite unfortunate. Be careful when you stand your ground meaning pick your arguments carefully. Arguing too often and your unplanned departure from the organization you were hired by won't be missed. I've always valued St. Dominic's (1170–1221 AD) words "Blows may avail where blessings and gentleness have been powerless." Another way of saying this is always being nice to everyone doesn't work well and sometimes you have to stand your ground.

Following your company's policies is extremely important. When you work for a larger company, the legal department has usually drawn up these policies because at some point in time, someone has abused these privileges. No alcohol during company time, conflicts of interests such as owning a business that you send work to, or adding non-company-related billings to your expense account are examples. I've seen these all violated by someone at some point and it has always ended badly for the person who was responsible.

Social and business networking services as well as the written word are highly used and today anything you write is available to your company or prospective company. Here's an example of what a written document that you thought was private can do.

When I was heading up a group of engineers someone who had previously worked for the company wanted to be rehired. Four colleagues were asked about him and their responses were noncommittal. Looking through the files to see the type work he had done when he worked for the company located a copy of a rather unpleasant letter he had written to senior management. It was about his displeasure with his former manager. He felt this manager was the reason he was not being promoted. I discussed this with the manager and discovered the true cause. Due to this information, he was not hired back even though he was a proficient engineer. One 8-year-old letter ruined his chances for a fine opportunity.

Problem solving meaning determining why a major piece of machinery or equipment failed and keeping it from happening again is serious business especially when a failure is costing a plant many thousands of dollars a day. Plant Manager's careers have been jeopardized by too many of these type incidents occurring on their watch. So when such failures occur, it's a good idea for the investigators to walk past the Plant Manager's office with a serious composure, shaking ones head from side to side, and looking down as if you are in the midst of deep thoughts about the failure. Why is this brought up? Well this is totally opposite of what an engineer feels when involved in such an investigation. Indeed, these type investigations can be quite exciting to work on and it is much like a detective gathering facts to solve a crime. It's not unusual to see two engineers assigned to the investigation walking past the manager's office saying, "This is a great failure," meaning of course that it's a failure that will require considerable technical resources to analyze and solve. Senior management won't share such enthusiasm on high visibility problems, so be cautioned.

2.3 DISENCHANTMENT WITH YOUR JOB

During my career, many people have come to me with statements such as, "I'm bored and nobody appreciates my work, I'm going somewhere else!" This is usually followed with, "What do you think?" This is a difficult question to answer since you never know the whole story. The following approach can help them. Tell them that when they have some time, sit down and think. They should get a piece of paper and make two columns. One column should be labeled "What's Good About It?" and the other column titled "What's Bad About It?" Explain that they should be truthful about the list since no one else needs to see it. It's been my experience over the years that the "What's Good About It?" side has more than 10 times as many items as the "What's Bad About It?" side.

The exception is if there is one item on the "What's Bad About It?" side that is so major that it overrides everything else. When this technique is used on personal issues such as buying a motorcycle, "My wife said No!" might be such an item.

A case history is always much more valuable than trying to explain the method, so let's look at one. This technique works on a lot of things: job moves, house changes, school choices, and so on.

Here's the process with a fictitious name but true story. It helped many people by letting them make their own decisions.

Charles had 23 years with a company and was 53 years old and said nobody appreciated what he was doing. He was looking for another job with another company. He went through the chart method and made up the following table:

What's Good About This Job?	What's Bad About This Job?
I have a job.	Nobody appreciates my work.
The salary is good.	
The health benefits are good.	
The saving plan is good.	
I am good at what I do.	
I am well known in the company.	
I can afford where I live.	
My family is happy.	
I like the community.	
I have a short commute to work.	
The work is interesting.	
I can retire with 75% benefits in 6 years.	
I have 23 years in the retirement plan.	
The company is solid and so is my job.	
I still get raises every year.	

Actually, the Good side list goes on and on. You might start to question yourself on who cares if you're not appreciated.

Consider if not being appreciated is really true since you're getting yearly raises. Many employees feel they aren't appreciated for all they do. Sometimes, it's true and other times you're not being realistic. Think of what you've generated for the company yearly compared to double your compensation since that is what you are costing the company. In any case, is it really worth giving up all the good things?

In this case, the fellow decided he had a good thing where he was. The opportunity for a job change within the company came up in a few months after he went through this exercise. He stayed with the company until he was 60 years old and he retired quite comfortably.

Unfortunately, this method doesn't always work. On my first visit to Japan, I was under the assumption that all Japanese engineers were very satisfied with their jobs because that is what I had read.

A young engineer I was working with asked me out one night. He said he had things he wanted to discuss with me about proceeding on for an advanced degree.

It seems he was quite dissatisfied with his work environment since an advancement was a long time in coming. He was considering going back to school for his doctorate.

We made up a what's good and what's bad about it chart. The conclusion was that it seemed worthwhile for him to discuss his concerns with his superiors and wait a year and see what might happen. He really had nothing to lose since he was thinking about leaving anyway. He was an excellent engineer and they probably didn't want to lose him.

Later it was learned that the engineer resigned a month later and went off to get his doctorate. The talks with his management probably went poorly.

Many times things don't go well at the office for us on a given day. We may come home as the above folks did mumbling about our work life.

You may be troubled by the following:

- Someone you manage isn't following what you told them to do.
- Your management is not following your recommendations.
- Your promotion fell through or your raise was too low.
- Your analysis just isn't working or no one is interested.
- You've gotten a task to do that you're just not interested in.

So what do you do about it? Well you can feel sorry for yourself and that sometimes feels good, but it doesn't really help the situation. You can tell a friend or family member about your problem, but that doesn't solve the problem. Sometimes, you just have to write down what's troubling you and look at it in a logical manner. After all, that's what we have all been trained to do as engineers and managers. There's always a solution to your problems, and once you admit to what the problem is, you can focus on what you're going to do to address them. Here's a hypothetical story that can help with a way to address such a problem.

Your Problem Is

- John was supposed to determine why the compressor failed and discuss it with you first. He took it right to the Unit Supervisor who didn't appreciate what John had said or the method he had used and confronted you about it on Friday.

What You Can Do About It

- Discuss this with John and tell him why he needed to discuss it with you first. Tell him this is the second time he has done this and it will be included in his performance review. It could have resulted in a safety incident and it is a reminder of what to do next time. Tell John this doesn't only make him look unprofessional, but it reflects back negatively on his colleagues. He should know that he will not be trusted with this type work until he can prove himself worthy.
- Talk with the Unit Supervisor informing him that this shouldn't have occurred. Mention that you should have made sure the work was reviewed before it was presented to him so in that regard it was your error too.

After writing this down as a reminder of what needs to be done on Monday, you can relax and enjoy the weekend with your family. Is it the perfect solution? Probably not, but it would have kept you from getting stressed out and having your blood pressure rise too high. It isn't worth ruining your health over.

There was an old movie with the title "The Seven Year Itch" starring Marilyn Monroe back in 1955. It was about a fellow married for 7 years fantasizing about this gorgeous girl who lived upstairs. A funny movie but not the 7 year itch I'm talking about.

Reviewing back the last 48 years, it seems that every 7 years I had this itch to do something new. When my job, meaning failures or analysis in my case, had the same type work coming around for a third or fourth time I was ready to do something different. I was lucky because my wife had the same itch. It has proven to be a good thing. It didn't necessarily mean leaving the company I was working for, just the type work that I was doing. Sometimes this meant moving to a new, bigger home, receiving a higher salary, and a new interesting place in which to live. These might be some of the reasons my wife didn't mind.

I'm sure many of you have similar feelings and if you do here are a few suggestions based on things I've done that kept work and life interesting. Obviously, leaving plenty of time for your family will keep you quite busy too.

1. I went on for my Master's degree at night in order to gain experience in other areas. With my Master's I could later get my Doctorate and then research and teaching were new options to explore. Remember the company you are employed with may not recognize your new degree and might have you doing the same thing. I moved to new industries after each degree for this reason. However, you may be able to negotiate something different, possibly even suggesting a new position for yourself in the company.

2. I was always aware of other positions that might be available within the company. By developing my niche in analytical modeling, I found it relatively

easy to switch from machinery to structures and fixed equipment thus staying diversified. Niches are important.

3. When an opportunity is available that no one else seems to want you might want to take advantage of it as it may lead to something new. Don't turn away positions as a supervisor or manager as there are valuable learnings and career paths in these areas.

4. Don't become too much of an expert in one specialized area as you might find yourself locked into that position. Your company will need you there.

5. Go to technical society meetings, be on committees, and present technical papers. Meet people in other companies and network with them. You will get a good idea of what others are doing.

6. Have hobbies at home too that will get you over dry spots. I liked working on and flying my airplane, amateur radio, restoring old cars, and manufacturing things.

7. Be aware of what your worth is in industry and be aware of other opportunities within your company and even outside of it.

8. Keep yourself trained in new technologies. I was always taking courses in the finite element method and other new piping and fracture mechanics programs. The more you know the better prepared you are for whatever comes up. Don't forget those communication and problem-solving seminars you should attend.

9. When you have a specialty that others might be interested in, present a company-wide seminar on it. I did them on mechanical engineering techniques. You might have some other specialty or course you've been to.

10. Don't be too anxious to leave your present company. Opportunities may come up that you didn't know existed. Think things out very carefully before you make your decision as downturns could be temporary.

2.4 CONDUCTING YOURSELF IN A MEETING

Correctly communicating to management what has occurred and what needs to be done is so important to an engineer. It would be wonderful if engineers had the verbal ability of attorneys in presenting data to management. An attorney's job is to make juries feel comfortable with what they are telling them, the decision that has to be made with the evidence and data they have. Unfortunately, many of us don't have this type of training. Fortunately, it can be learned by experience and watching other successful engineers. Your company's senior technical personnel didn't get where they are via a lack of communication skills or poor judgment. There are three things that are good to know when discussing work with senior management. The first two items are self-explanatory, but the last will require an example:

- Management does not like to hear bad news, so present positive plans.
- Management does not want to hear a wish list of solutions, so present only your best and most cost-effective choice.

- Management is not impressed with complex analysis or technical terms, so the engineer must simplify the cause, solution, and implementation so it can be understood and acted on. The management you are presenting to may not be familiar with mechanical engineering and may have their expertise elsewhere. It is useful to adjust what you are presenting to suit your audience.

We have all been in meetings, where the engineer droned on about a problem in technical babble to an audience that wasn't technical. We were embarrassed for the presenter and wished we could have told him or her to talk in terms the audience can understand. The presenter many times is so excited about the great technical job the team has done that the major point is missed. All that the management team wants to know is what was the problem, was it corrected or how are we going to correct it, how much will it cost, and how long will it take.

If the presenter can provide simple explanations and graphics explaining the technical part, this is usually appreciated.

Suppose you are discussing the resonance of a structure and its failure. Now as engineers we know resonance can be a highly damaging vibration caused by exciting a structure's natural frequencies. Resonance could be clearly demonstrated by bringing a tuning fork to the meeting striking it and showing the resonance of the tuning fork. With no continued exciting input, meaning striking it, the vibration dies out. However, with a continued input and without material damping mention that the tuning fork would eventually fail in fatigue.

The fatigue failure part could be demonstrated by bending a paper clip back and forth to demonstrate and explain fatigue. At any point you can stop bending, but some of its life has been used up because a very small crack has started to grow with each bend.

When reviewing a technical presentation, it's wise to make a list of the questions you think may be asked during the meeting and research them thoroughly. This allows the best and shortest possible answers for the audience you're presenting to.

There are several type personalities you may become acquainted with in meetings. Being aware of them will allow you to better understand them.

A pontificator of the obvious is one type that a colleague of mine named. This type person will repeat back to you exactly what you have just stated. It might be something like, "Let me understand what you just said" and repeats exactly what you said. This happens when the person has nothing creative to add but is at a high enough level that he or she feels they have to say something. They contribute little to a meeting, but you will need to be patient with them.

Another colleague mentioned that at a meeting he had attended, it was a Bastion of Bobble Dom. When he was at the meeting everyone was smiling and nodding their heads, but no one really understood because no one said anything. Remember those figures whose heads bounced around on car dashboards? This is mentioned only so you will remember not to just sit at a meeting. Have an idea before the meeting on what you want to contribute and when you have something worthwhile to say think it out and say it.

Memorable managers are those who would sit quietly in a complex meeting taking notes as the various people provided input. At the end of the meeting, the manager would outline a plan forward using the input provided. Everyone left understanding what they needed to do and feeling their input had been recognized.

When collecting data at meetings, make sure everyone's input is considered. Many are worried about saying something wrong and looking unprofessional. One way to overcome this is to write all the suggestions on the board no matter how inappropriate it may seem to the group. The approach is then to say that anyone can withdraw their suggestion if they feel other data makes it questionable. So if someone says "I think it failed because it was painted yellow" and finds out it doesn't matter it can be taken off the list. The person will be respected for recognizing this. Telling a participant "That's a ridiculous idea!" and you probably would not get another piece of information from him or her. Who knows the next words out of the participant might be the solution to the problem. So at meetings draw even the quiet participants into the conversation. For example, "Susan, I remember you had a similar failure at your location. Can you tell us a little about it?" This usually works well.

Meetings with too many attendees can become unruly and much input may be lost because of all the cross talk. Calling in small groups at a time to present their data is one way to address this problem.

2.5   ORGANIZE AND PRIORITIZE

Everyone should have a plan that sets their priorities in life. Priorities can be worked every day. A plan, on the other hand, goes into the future. Your long-term work plan might be to work designing machines of some type with a long-term plan to become a Chief Engineer at some company. However, your work plans and your life plans should mesh but will be different. For example, if your job will take you away from home most of the year and you have a family of four they may not mesh and you will have to make choices.

This reminds me of a management seminar I attended while trying the managerial ladder for a couple of years. It took me that long to realize I was much better equipped to be a technical guy.

There were about seven of us in the class and we were asked to write down the four most important things to each of us. The other managers' lists went something like this, in various order:

1. Safety
2. Profits
3. Productivity
4. Development.

I went to the board last and mine went exactly like this:

1. God

2. Family
3. Friends
4. Work.

Now this didn't go over well with the other managers who all said I had missed the point of the exercise. I said I believe they had all missed the point. This all came back to haunt me later and thus my rather short stint as a manager. This is not a recommended approach if you are trying to impress others for your future managerial career development.

The point of course is to remember that there is a lot to life and the work environment is only one portion but an important one.

As for prioritizing, it's important to keep a list of what you are working on, when it's due, along with its urgency. We are all usually juggling several projects at once. Unfortunately, we tend to spend more attention to the ones that are interesting. There is the drudge work that just has to be done, which can usually be set aside for a certain time. Responding to all the various networking calls might take place later in the day. We are all most productive and creative in our thinking at a certain time of day. For me, it was the morning and I would come to work quite early before anyone else arrived and do that type work. It might have included an analysis of some type or a new design. We all have different schedules and it's important to determine yours so you can prioritize your jobs. So as the title of this section suggests, organize and prioritize your life and your work.

As to urgency, it's important to remember those given to you by your immediate supervisor. When he or she says something is important, you need to realize they should be of high priority even if you don't think they warrant the extra effort. We never know the whole story and there could be some hidden agenda we aren't aware of. This is the person who should be looking out for you and your future. You should therefore be as helpful as you can.

There are times when you are working on a critical job and your supervisor has work he or she wants done. The proper thing to do is just ask if you should stop the job you're working on to fit the new job in. That way, jobs that would affect your supervisor and that he or she may not be aware of can be addressed.

2.6 DO AS MUCH AS YOU CAN FOR YOUR COLLEAGUES

As mentioned earlier, colleagues were very important to me and when supervising people, I felt it was my duty to consider their careers too. The Golden Rule "Do Unto Others As You Would Have Them Do Unto You" is a good rule to follow.

I kept a record of the good work they had done along with the savings to the company. They wrote reports and documented their work. They presented their work before management as a presentation that I had reviewed. When someone else took credit for work done by them I reacted quite aggressively and would discuss it with the perpetrator's supervisor. It didn't happen often. Notes were also kept on what others had done, so I could compare their work with my group's. In this way, when the budgets were being prepared and raises and promotions were to be given, I'd be prepared at the meeting. Once someone commented, "Here he comes with his book." The book of course contained defensive data for my group. One of my promotions was probably because the other supervisors didn't want me in that meeting anymore. When you do right by others, most will work hard and be loyal to you.

One member of my group had the responsibility to expedite jobs. He did this better than anyone. He had a wonderful way with people and would go right to the top of a company's management chain to get a job completed for our company. At one point, when I wanted to see what Rich did, I went with him to follow a job at a large fabrication shop. Someone came up to me and asked what Rich's title was. I told them that he was the Vice President of Outside Repairs. He responded with, "We thought he was important because when he comes here he talks directly to the Plant Manager to get results!" After that tour, there was never a concern about how Rich did his job again.

One time while doing an analysis, a technical paper that was in German was needed and I couldn't find it anywhere. Rich stopped by and the next day he came back to me with the paper translated from German. That man had contacts and could get results.

2.7 THE CATCH 22 OF ENGINEERING PROJECT WORK

Finding the causes of problems and fixing them so they wouldn't fail was my occupation for many years. The Front End of a project is the very beginning. For example, if you were building a one billion dollar petrochemical plant, the Front End would be at the start before any equipment was designed or purchased. The land might just be starting to be cleared. This is the time you want to make improvements in designs, so there is minimum cost impact. You can learn from past inadequacies and implement them now. Troubleshooting occurs after the plant is up and operating and the inevitable problems are found. If there aren't small problems, you've probably wasted a lot of money on over design. It's the big plant shut down or safety problems you want to avoid.

This is where after much experience I can now write about Catch 22. In case you've never heard of Catch 22 or read the 1961 novel Catch 22 by Joseph Heller, let me explain this lose–lose phrase of circular logic as he did in his book.

- You can only be excused from flying bombing missions on the grounds of insanity.
- You must request to be excused.
- If you request you are in fear for your life then you must be sane since an insane person would not make the same request.
- Since you are sane you are not excused but you can request to be excused.

With the definition understood back to the Catch 22 of Troubleshooting.

1. When you do a detailed analysis and find a design defect and remedy it after a plant has started up the response from management is "Why didn't you do that on the front end of the project?"

2. When you have done something good on the front end of a project no failures happen, so management feels they don't need any front end help on the next project.

3. When the next major project comes along funding for the front end work is cut drastically, so problems happen and then they figure they need more front end input for the next project. Thus the never ending loop.

After this happens several times, you would think someone would become the wiser since it's all documented. It doesn't happen that way because project costs always rule supreme.

2.8   ARROGANCE, HUMILITY, FAVORS, AND COURTESIES

At some point in our lives when we are very good at something and become well known, we may be known as experts. Some say someone is knowledgeable if they are aware of a technology. They are an expert if they have actually used the technology. I'm being facetious but here are some other examples.

I thought myself an expert in a certain area of engineering. Papers had been published and many problems solved and I felt pretty good about myself. I had become active in a technical society and every year took my wife with me. Cruz can't stand to see the wives from other countries just sitting around, so every year she would collect the wives from Japan, Germany, France, United Kingdom, or wherever, and go on a tour in the City we were in. She would see them every year. One year I had developed what I thought was the best technical paper I had ever written and expected this was a paper that would receive an award at this particular meeting. I was at the point of being arrogant. After the session, as I came strutting out I was greeted by the Society President whom I felt certain was going to congratulate me. Instead he said, "Aren't you Tony Sofronas?" I told him yes and waited for him to comment on the greatness of my paper. He remarked to me, "My Wife Knows Your Wife Cruz." So went my fleeting moment of stardom and another good lesson for me in humility.

Most engineers when they are good at what they do tend to be a little arrogant. That's not a bad thing as long as it's not carried too far. Engineers and most people in general can be just plain annoying when too arrogant. Saying that you accomplished something with no help from anyone is arrogance. Someone might not say this directly, but if you do work and presented it as your own with no recognition of others this is what you are doing. A highly technical paper with no references is an indicator. As repeatedly mentioned in this book, no one really does anything without the help of others. Some of my biggest successes were with help from others.

Arrogance is also apparent when a person expounds on their accomplishments. It's important to be proud of what you do or where you have been. Most people especially your colleagues, family, and friends will listen politely. This stops being arrogant when you are truly interested in what they have done and where they have been and ask them about this. Others have accomplishments that are probably far superior to yours and it is appropriate to ask them about these too. While you may be an excellent engineer and have traveled the world, they may have raised a family they are very proud of. It's nice to ask and know.

One can't help but be humble in engineering as there is so much to know and you can only know a small fraction of it. Tomorrow the fraction you have just learned may be obsolete. Someone called me an "expert" on gas-engine compressors. I was embarrassed because there were machinists and technicians who knew much more on the operation, troubleshooting, and repair of these machines. You may know one small part of a subject well, but the term expert will have to be better defined before I would ever want to use it or be called one.

The term Favors isn't discussed much, but it is present in industry. Here it will be defined as when you help someone out usually getting them out of a difficult situation. It's done more as a courtesy and you don't expect to be repaid in any way. It was just the right thing to do at the time. They are important and when I retired and colleagues asked me why I was retiring I use to say, "Because all the people who owed me favors have retired!" That wasn't really true but what that really meant was that most of the people I could depend on had all retired.

Here's an example. I was known for understanding gas-engine compressors and had seen many problems with them. A Supervisor called me from a gas plant in Louisiana and he asked if I could come down to talk with him about a problem he was having. It seems that at this remote site a large gas-engine compressor had a crankcase explosion and they didn't know why. Engines such as this have explosion relief doors so nothing really breaks, but it is quite frightening and can be dangerous. There is a loud explosive noise and large flames may be seen emanating from the doors especially in the dark. They were afraid to restart the engine as careers were on the line since a high-level manager would be witnessing the start-up. After several days and measurements, a very experienced technician in my group and I identified the problem as a well hidden loose internal part. Actually, Don performed all the hard work such as crawling around inside the engine hitting bolts with a hammer and placing and reading dial indicators. We determined that when the engine was operating, a loose power cylinder would allow the piston skirt to rub as shown in Figure 2.1 and the heat generated would ignite the oil fumes in the crankcase causing an explosion. If they would have started up the engine, it would most likely have happened again. The Supervisor was extremely grateful since his superiors saw him as a problem solver since he called in the right people. Thus, a favor had been done for him and he repaid this favor by calling my group in for other jobs.

I also owed people favors. When someone recommended me for a job or did something for me that helped me look good, then I owed them favors.

Most of the people who reported to me did me favors by doing their jobs well and making my group a successful one. I recognized them for this in their performance reviews, raises, and promotions.

It's not something that is recorded in a tally book such as he or she owes two favors and you owe them three, but it's just something you feel comfortable doing.

Many times it has nothing to do with favors and is more just plain courtesy. Very often as a consultant I'm asked for information or advice, which I am happy to provide pro gratis. About 50% or more of the time after providing the data to the person they

Figure 2.1   The piston rub.

never reply, not even a thank you. Other consultants say they have experienced this also. The purpose of bringing attention to the issue is that this lack of courtesy will close a lot of doors that normally might be very helpful. Not returning telephone messages also falls into this category.

How you conduct yourself does make a lasting impression on people. When something reflects badly on you, it also reflects on your company and this can get back to your management.

2.9   BE CURIOUS AND INQUISITIVE

Engineers tend to be curious and inquisitive people. It's in our nature and is probably why we wanted to become engineers.

My first entrance into working on machinery was when I was about 13 years old and was rebuilding an old Model A tractor that my Uncle Pete and Uncle Jim built in the 1930s [2]. They built it to pull heavy hay wagons because they felt sorry for the horses in the hot summers. It had been totally disassembled when I found it 30 years later. The engine was half buried in the ground and the rest of the parts were spread all over like a debris field.

I however had a restoration plan, even though I knew nothing about machines. My plan was this:

1. Collect as many parts as possible from the debris field and hope they were all parts for the tractor.
2. Dig up the engine and find any place that looked like parts should be bolted on.
3. Find any connections that looked like they should be connected with hoses and connect them together.
4. Find anything that had the same colored wires and looked like they should be connected and connect them.
5. Obtain nuts and bolts from my Uncle's antique cars and use them on the tractor.
6. When everything was connected up, borrow the battery from Dad's car along with draining some gasoline and give it a try.

After two summer vacations, the time of reckoning came. I hit the starter and to my amazement, it cranked over. It wouldn't start though. Funny thing was that the gasoline was pouring out of this big hole. I figured that big hole had to be covered so the gas wouldn't come out. By now all the fuel had poured onto the ground.

My Uncle Pete stopped by and wandered over to see how I was doing. He looked at it and then said in a calm voice, "Tony I think you better get a carburetor for this thing," and walked away. Carburetor, what the heck's a carburetor? I took a look at a book that said the carburetor is to an automobile engine what the heart is to a human body. I may not have known much about engines but I did know I needed my heart, so a carburetor must be important.

What I had done was piped the fuel line directly to the windshield wiper vacuum connection since the two fittings looked about the same. Someone had probably taken the carburetor for use on something else many years ago. I saved my money and bought a Zenith carburetor that fitted the hole perfectly. It started up and as I drove it through the hay field it would start little fires here and there from the fire coming out of the straight exhaust. My Uncle Pete suggested it needed a muffler. A muffler what's that? I thought a muffler was something you wore around your neck to keep warm in the winter.

My sister drew the following sketch many years ago. It is positioned over my computer and I look at it often (Figure 2.2).

Being curious is a good thing because it's a way to learn. Here are some suggestions based on my experiences.

In my case, it was a way to gain knowledge when I was an engineer. I would look into how other people solved problems and would store this for later use. Much time was spent in machinery repair shops following jobs to see what was being done. While in the shops questions would be asked about repairs being made on other equipment and I would look at the type failure and how it was being repaired. This was useful as similar failures were usually observed later. It was nice to have seen it happen before and to know what was done.

One company I was employed with had an excellent metallurgical laboratory right down from my office. I would stop by each morning and visit with the technicians and discuss what they were working on. Since my company had plants around the world, there could be failed parts from the United States, Germany, Japan, France, or

Figure 2.2   Working on the Model a tractor.

elsewhere. My responsibilities were worldwide, so it was an excellent opportunity to see what was going on. When a large gear, turbine disk with missing or damage blades or large failed bearings were observed, the plant would be contacted to see if my group could be of help. While this might seem like an intrusion on someone's territory, it really wasn't. Metallurgy and mechanical engineering work well together. While the laboratory can tell what type of mechanical failure has occurred they usually don't know why, with the information they have received. A mechanical analysis may be necessary to define the cause and to determine the method of repair. For example, the laboratory might say the failure was due to a bending fatigue failure in a sharp radius and that the radius should be made larger. A mechanical analysis might say the failure was due to "V" belts that were over tightened by an inexperienced machinist. This caused a bending fatigue failure in the fillet. Training may be suggested and no redesign to the shaft as the fillet radius was correctly sized.

I'm usually seen doing some type of calculation. It's like training for a sport. Unless you keep at it, you will never improve. Without developing your skills, they may not be sharp when you need to use them.

Many times I may come across situations that are interesting and need an answer. Section 10.12 on the Shuttle Columbia accident is an example. There was no request for me to perform this analysis, but statements made were unbelievable. I did this work to prove a point. Experiments conducted later prove my analysis correct and that was gratifying.

The same was true for the truck crashing into the elevated compressor platform column support in Section 10.14. I happened to see the accident after it occurred. The plant had concern there was possible damage to the compressor and other equipment on the platform. I volunteered to do the analysis to help out.

The literature available online is plentiful and valuable. It's amazing how many questions I received as a consultant, whose answers are readily available online but have not been researched by the questioner. As an engineer, one needs to become proficient on searching the literature. Even out-of-print books and articles are available from libraries such as Linda Hall Library in Kansas City, Missouri, which were utilized when researching several areas.

Research can come from very strange sources. In Chapter 10, analytical models are discussed and the need to obtain verification data on the model is shown since the analytical model may never have been used before. In most cases, such test data is very difficult to obtain. In Section 10.10, on the dangers of pneumatic testing, a viable piece of information was obtained after seeing a program on a contest where pumpkins were shot out of a cannon using air pressure. The distance traveled for one 10-lb pumpkin with a column of 90 lb/in.^{2} pressurized air behind it was over 3,800 feet. When the analytical model was applied this test data was in the range of the calculation method. When a pressurized vessel explodes it's not pumpkins that fly that can injure personnel but metal pieces; however, the principle is the same.

Online courses and example problems are extremely useful. I had a finite element analysis program (FEA) for consulting work and used it for stress analysis and heat transfer problems. The program also was able to perform computational fluid dynamics (CFD), which is useful for analyzing fluid flow problems. I had been involved with FEA for many years and learned CFD from online courses and examples. It was very cost-effective and I used it on one of my consulting jobs. The client was quite pleased as the analysis was simple enough to show colored flow profiles on what was occurring to her management.

2.10   STRIVING FOR PERFECTION

Perfection is defined as, "The state of being flawless." This is a wonderful goal to strive for in faith and marriage, family, and friends. It is something we can work on throughout our lives. Reading, listening, watching others, failing, and trying to do better is how we strive for perfection. Of course, we can never be perfect, but we try to get as close as we can.

In the field of engineering if I had sought perfection in the first analysis I had performed, I would have been fired. That's because after 48 years I'd still be working on it. I would have been fired because I wasn't getting enough work done to justify my worth to the company.

Obviously, perfection isn't something we want to attain when doing design work or troubleshooting equipment. The goal in design work is to design to the customer's specifications and no more. Trying to attain perfection will make the product unreasonably costly and unable to meet the production schedules. Likewise, when troubleshooting or problem solving, the task is to get the equipment back into production safely, with adequate reliability and in the quickest way possible.

Does this mean we shouldn't try to do the best we possibly can? Certainly not, but it does mean we should be aware of the time restraints.

My niche, analytical modeling helps in being timely with solutions. The analysis is kept simple and can quickly provide answers. The answers may not be as accurate as I would like them to be, but they are enough to explain problems in detail and propose possible solutions. You will see more of this in Chapter 10.

Most of our activities in engineering require some sort of resolution in typically less than a month. Failure analysis, design concepts, decisions on equipment selection, and review of new products or projects have schedules and there is little time for trying to attain perfection. An adequate, timely, reasonable, and believable result is all that is required.

An engineer will not progress far if management feels the engineer cannot be depended on to get the job done in time. Good engineers are capable of working several projects at one time and having them completed in a timely manner. Proper prioritizing allows them to do this. In my case, many of the complex analysis I have done are very time consuming and were done at home at night or on weekends on my own time. Not because I had to but because I enjoyed doing them and felt they were necessary for making an informed decision.

Perfection is a wonderful goal to strive for when performing a task, but we must be satisfied with only achieving somewhat less than perfection if we want to be successful.

REFERENCES

1. King, W.J., The Unwritten Laws of Engineering, A.S.M.E, New York, 1940.
2. Sofronas, T., Family and Friends in the Oil Patch, EP Press, 2004.

3

UTILIZING THE INPUT FROM OTHERS

3.1 JUST OUT OF COLLEGE

Once there was a complex problem that had to be solved. The working hands all had a good idea of what needed to be done and all had reviewed the job. I had just been hired right out of college and was being introduced to the machinists, operators, and crew supervisor. My manager was introducing them to me and said, "Tony is an engineer with a degree in mechanical engineering and is ready to help." With that the crew supervisor said, "Well that makes a big difference! I guess we had better start explaining the problem to him with the basics."

Things didn't go smoothly after this. A new locomotive engine was being tested and we were in the test area witnessing the start-up. Being a new engineer in the early 1970s, I was dressed in the decor of the day, a new light suit, white shirt, and tie. As the engine was being started, I noticed that everyone was on the other side of the engine. A burly fellow named Marv who was the Test Cell Superintendent was signaling me to come over by him. I decided not to follow his suggestion as I had a much better view of the engine from where I was standing. It was a little strange that the side of the floor I was standing on was black and dirty and the side they were standing on was a clean gray. I figured he wanted me on the other side so I wouldn't get my shoes dirty. The engine started up, then the oil hose broke, and sprayed the front of my suit. Lube oil doesn't come out of clothes. A few days later, Marv stopped by to comfort me and in his pleasant smooth Southern accent said, "Son, whenever a piece of equipment is being started up or is running always stand behind the most

Survival Techniques for the Practicing Engineer, First Edition. Anthony Sofronas.
© 2016 John Wiley & Sons, Inc. Published 2016 by John Wiley & Sons, Inc.

experienced person at the site." Another old hand told me a different version of this many years later at a refinery. Arlon said, "When there is a fire in a petrochemical plant always follow the machinists, never follow the operators. The machinists are trying to get away from the fire but the operators are running to the fire so that they can shut down valves and equipment."

In your early years, it's wise to listen to the advice of experience and not to assume you know it all.

3.2   MENTORS AND COLLEAGUES

Many of us are lucky to have good mentors early in our careers. I tried to follow my father in being a good family man and having good moral and work ethics. Since he wasn't an engineer, I had a technical mentor in industry when I started my career. This fellow was an excellent engineer, well respected, and analytically one of the best I have met. His analysis was extremely exact with his mastery of the sciences and his solutions were always unique meaning new and not just copied.

The company had a 1-year course, where a selected group of us were to use analytical methods to solve actual company problems in heat transfer, electrical circuits, vibration analysis, fluid flow, stress analysis, and all branches of mathematics. We worked in teams with various disciplines on each team. I was the mechanical engineer and there were also two electrical engineers. My mentor was the course director. We worked problems the company was involved in after reviewing the appropriate advanced theory and were expected to use the theory learned to solve the problem. New motor designs, cooling of electronic packages on space satellites, and braking resistors for locomotives were worked on just to name a few. It was a wonderful learning experience and a great way to get indoctrinated into an engineering company. Much of my passion for engineering came from this experience and from the input and abilities of my mentor who never realized he was my mentor.

Marty was a very analytical Mechanical Engineer and used his abilities to solve practical design problems.

An analysis by him usually contained calculus, differential equations, imaginary numbers, and other mathematical treatments, used in the most eloquent ways. He was well known throughout the company as being able to handle the difficult problems and putting a practical solution on them. He would go out of his way to find these problems. He eventually became Chief Engineer, a very prestigious position in this large company.

Very rarely do you find someone like this and when I would do an analysis that he didn't quite agree with, I would see a note on my work such as, "You had better charge up your calculator batteries you were sloppy on this one."

When he got bored he went around looking for difficult problems that no one else could solve and he solved them.

Now there are also poor mentors also so one has to be aware of this. Luckily, I never had one but have seen them in action with other less fortunate engineers. These people usually don't really want to help you and to them you are just a burden with your questions. They feel the information they have developed is for their advancement only and are not willing to share it. They become known for this and young engineers are told to be aware of their traits. It's not a good reputation to have and what's most unfortunate is that some of these folks are very good at what they do.

Not much time will be spent on this because I usually tried to distance myself from these negative types and have erased them from my memory. Not so with my mentors as I will always remember them fondly. Much of their thinking is in this book and in my soul. Remember that when you see something you do not like in a person that's your signal to do things differently.

3.3   INTERACTION BETWEEN DISCIPLINES

When troubleshooting, we can become fixated on the discipline we are most familiar with, for example, mechanical engineering. Investigating the failure of a machine might have several different specialists involved in the troubleshooting effort. When one's specialty is vibration analysis the first thought might be a torsional or linear vibration problem. A stress analyst will be hard at work trying to develop an analytical model of some sort to describe the failure. The materials engineer will be looking at material properties and corrosion and the process and control engineers will be reviewing computer simulations and the possibilities of abnormal operations.

It's human nature to want to analyze problems with the tools you are most familiar with.

When the solution is obvious, the one discipline approach can work. A specialist may have seen a very similar failure and can go right to work solving it. Indeed, most day-to-day failures are solved by replacement of parts or with the field personnel applying a simple fix. Replacing damaged bearings because the oil was contaminated and correcting why it was contaminated will mitigate the problem.

This doesn't work well on complex multivariable system problems. Erroneous solutions can result in severe consequences to equipment, personnel, and careers with the one discipline approach.

This can be addressed by using a structured team problem-solving approach with all the necessary disciplines represented and a team leader who has been selected by management. In this way, all of the potential causes can be identified using the talents of many.

There is no purpose in outlining all technical problem-solving methods as they are well documented [1, 2]. Most of us have taken a seminar or been involved in such sessions and know the benefits.

What I have noticed is that there are times these methods have not been applied but should have been. There have been reasons for this such as given below:

- A deadline is approaching and it is imperative to careers that it is met so the uninformed make the critical decisions.
- No one knew that such an approach would help and no one had been trained.

- The problem solving is well underway before it is realized that the approach is rather chaotic and more structure is needed.
- There is not much urgency in solving the problem, meaning no one wants to spend any money, delegate personnel, or be responsible.
- The solution to the problem was started with no action plans and no definite responsibilities or the various disciplines needed to solve the problem are not available at the site.
- There is one dominant high position member of the team who drives the solution his or her way.

All of these reasons are management's job to address. The question then is when is it necessary to utilize a structured team approach to solve a problem?

Such methods are beneficial when safety, litigation concerns, major production losses, or careers may be at risk. When the organization may be the cause, a formal approach would be required. There have been excellent articles written on when just making a repair without thought has been inadequate [3] and more should have been done. A quick Internet search on Engineering Disasters will show many such failures and that most were avoidable. Reading these articles are reasons to consider structured team problem-solving approaches and training in root cause analysis methods such as Kepner–Tregoe™ or TapRooT™.

With the litigious society we live in today, it is well worth investing in methods that produce a "paper trail" to show that a sincere and thorough effort was made to address the concerns.

3.4 IT'S NICE TO BE APPRECIATED

In the work environment, it always amazed me how jobs that were well done aren't recognized. It could be a good repair job by a machinist or someone who has developed a new design. Too often we have heard, "It's his or her job and that's what was expected!" At one time there was a program that would pay the factory worker a small amount for coming up with ideas that would save money. It was popular with the workers but unpopular with others who said, "They're just doing their jobs why should they be paid extra?" It didn't matter that the program had saved the company hundreds of thousands of dollars for a paltry commitment to the workers. Possibly it was felt that if too many good things were said to their employees they might expect a raise or a promotion. I certainly hope these weren't the reasons.

Once while working with a fellow after the idea program was canceled he mentioned a solution to a problem that was costing the company thousands of dollars a week. I asked him why he didn't say anything to his supervisor and his reply told it all, "Why should I. There's nothing in it for me and my supervisor will get all the credit!" I had him write up his idea and took him to present the idea to management. They accepted it and this made it possible for the fellow to head up a team to implement his suggestion. He did a great job and this meant a lot to him and he's been a contributor since. All it took was a little recognition.

It's good to compliment people on a job well done. Not because you want something in return only because it's the right thing to do. You probably can remember the names of all those who have complimented you over the years. The names of the ones that hadn't usually just fade from your memory. It matters to people.

One of the biggest compliments you can make to a person is to remember his or her name. This is a real tough thing to do if your memory isn't so good. I have gotten around this by putting the people on a 3×5 card and keep it folded in my wallet. What the person looks like, wife and children's names are all on the card. It helps but is not foolproof and there are still a lot of people called Hi, How Are You! All I have to do now is remember where I put the cards.

My wife once took a course on association that taught you how to associate a person's name with something funny. She knew a Mr Cook, who was a large, red-faced fellow, but kept forgetting his name. She pictured him with this big white cook's hat on, working in a kitchen. It worked well until the day she introduced him to her boss as Mr Baker. Association has its problems too.

3.5 THE FUNNY LOOK TEST

When I had been working on a detailed analysis and was unable to develop a solution, my mentor used to tell me, "Have yourself a cup of coffee and work on something else for a couple of days. Come back to what you were working on and it will appear much clearer." This was a good advice for anything complex you do when you get muddled down in the details.

This advice came up again many years later when a colleague was reviewing someone's work and said he didn't think it passed the "funny look test."

Here are some indications when something doesn't pass the "funny look test":

- When you tell somebody your idea and they gaze at you with a blank look. They don't have a clue as to what you're talking about.
- It doesn't look like something you have seen in the past and your instinct says no.
- It seems to violate one of the fundamental laws of nature. For example, your computer model results show that the tower you designed is bending in the opposite direction the wind is blowing.
- No one else believes you.

If any of the above occurs, it would be wise to review your results or ideas with a trusted colleague or friend who is also familiar with what you are trying to do.

3.6 UNCLUTTERED THINKING

I remember others who have said things or had done things that positively influenced me. They were usually people who had what I like to call, "Uncluttered Thinking."

They did or said things that made good sense and in your mind you would say to yourself "I should have said or done that!" It's hard to put into words and there are many stories in this book that you will recognize as this and here are some additional examples.

Example 1: When going through college I had a hard time with one course that was chemistry. The thoughts of getting a D were making me ill. One day I was told, "I don't recall ever having anyone die in my class because they received a D." I received a C, but this put everything in a new perspective for me and I use it when someone is frantic about a grade. Wish I had thought of it.

Example 2: I retired at 60 years for several reasons. I could because my wife and I had invested wisely, I wanted to spend more time with my wife, I wanted to consult, teach seminars, and write articles and books. The one thing that helped me make the decision was seeing a little framed stained glass message my mom had left me and it said, "Along the Way Take Time to Enjoy the Flowers." In this fast paced world, sometimes we forget to enjoy the things in life that really matter.

Example 3: Sometimes you require a slap on the side of your head by someone with an uncluttered mind. This is why I always welcome inputs from others on something I have done. I may not like what I hear, but I still think they are important. Once I was performing an analysis on an 8,000 horsepower extruder that was making plastic pellets out of raw product. I was having trouble visualizing how the product was processed and forced through the die. When I described the operation to my wife she said, "It sounds like you're grinding and making hamburger." That little piece of information cleared my mind from all the equations and back to a simpler uncluttered model.

Example 4: When developing analytical models, I can become so involved with the mathematics that I deviate from the problem I'm trying to solve. I find myself doing it just for the sake of developing a complete solution in the most eloquent way. Early in my career, I was told by an uncluttered mind, "Don't get too caught up in the analysis. Just solve the problem and get the equipment built and running."

Example 5: I remember hearing, "Sometimes you have to do something even if you're not sure it will work. You can always modify it but at least you have done something." While this is uncluttered thinking in problem solving, in engineering you have to be careful with this. In analyzing failures, the wrong conclusion and just trying something could be catastrophic. I wouldn't want to be on an aircraft at 36,000 ft, where the maintenance crew "just tried something."

Example 6: When I was learning to fly I was having trouble landing my aircraft and being right on the centerline. Instructors gave me hints and it all helped, but windy days were still a problem. Landings were always safe but just not as exacting as I wanted them to be. I flew with a pilot friend who had been a military pilot and he said, "Just put the centerline of the runway between your legs when landing and you will be at most 2 ft off." This uncluttered thinking did wonders and Bo was right on. The pilot seat was offset this much from the centerline and between my legs put the aircraft nose wheel at most 2 ft from the centerline.

Example 7: "Don't wait if you really want to do something, because you don't know what will happen in the future." I guess I said this to someone in one of my uncluttered moments. Of course, it has to be within reason. Don't quit your job and take a trip around the world with your life savings. What do you do if you live another 20 years?

3.7   THE ART OF VISUALIZATION

In this book, visualization will be defined as a memory or picture you have or develop in your mind that is helpful. The brain is capable of many amazing things and one of them is producing mental images. Some have this capability much more than most of us. Albert Einstein could visualize himself riding a beam of light to see what things looked like at that speed. Nikola Tesla could build and test machines in his mind and watch them operate for long periods of time. When they failed he would modify them in his mind. He would then describe the final design to a draftsman who would write it down and have it built and it usually worked [4].

Most of us have this ability to a much lesser degree. I believe we can develop this ability with practice. I don't receive wonderful colorful visualizations, but I do see myself within simple machines as a cartoon character watching what is occurring. I also see a simplistic version of the machine much as some of the sketches in this book. Figure 9.4 was imagined when I was reading a report on a broken crankshaft. Figure 10.16 with Sherlock noting the rubbing and Figure 10.17 of me pushing on the spring are examples of how I saw the models simplified. This helps when I am trying to build a mathematical model of the machine to explain a failure. I actually see parts rubbing together and heating up, vibrating or being impacted with a hammer. This ability only developed when I had taken apart many machines and had seen many failures. It wasn't an ability I had when I was young. That's why I feel it's a learned ability anyone can do. Geniuses have trained their extraordinary brains to do this much better than most of us can and therefore can develop exquisite theories. We can use it in less spectacular but quite helpful forms.

My life tends to be much more enjoyable because of this imagery. I can see things in ways not written down. A book I wrote for my colleagues and family was only possible because of these images [5]. It's a way of remembering historical events.

For example, once I observed a supervisor who was disciplining a machinist operating a lathe and he said to the machinist, "Roll up your sleeve or you will look like a barber pole." My mind visualized this as shown in Figure 3.1 and whenever I see or work on a lathe this image comes to mind and keeps me safe. I practiced trying to learn enough sketching so that I could explain situations like this.

There's another type of visualization. When someone is telling me something I don't agree with, I will try to visualize myself as the other person in order to see what they are thinking. There are volumes written on this subject; however, in its simplest

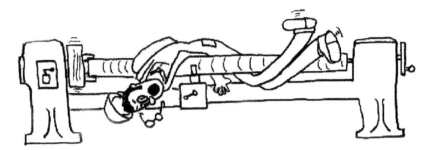

Figure 3.1   Roll up your sleeves.

form it's just "Putting yourself in the other person shoes." For example, if someone tells you that they don't agree with your analysis, if you are like me you will probably get quite defensive. A question to ask yourself might be "Why would I have said something like that to someone?" Some of the possibilities might be as follows:

- Your work might actually be in error, so you would want to clarify why they thought it was wrong and address it.
- They may like your work and be trying to help you improve on it.
- As discussed in Section 2.2, you may be invading someone else's territory.
- They might be doing a similar analysis and want theirs used.
- They may be jealous of your work.
- They may just have a disagreeable character.

Sometimes visualization is difficult when all you are seeing is red. Fortunately, most of the time, my friends were just giving me guidance. During my career, I have seen all of the situations mentioned earlier occur to others.

3.8   THE IMPORTANCE OF ALLIANCES AND NETWORKING

You may be the best and brightest engineer your company has ever had; however, if you are unknown this may all be in vain. I can't name many famous people who stayed to themselves, didn't publish their works, or didn't have many who knew of them.

One can't sit in their office producing work and be successful in engineering. You are known by who you are and what you do.

I have found that just writing articles, technical papers, and books are not enough to become well known in engineering. It requires contact with others to develop the needed networks. Some of these contacts can be extremely helpful such as a chief engineer, design engineer, or field service engineer for an Original Equipment Manufacturer (OEM) of equipment that has failed. With such alliances, others are available to help you with difficult problems or the need for data. This goes both ways as you will also be able to help others too.

Some of the alliances that have proven helpful to me are as follows:

- Engineers in OEM plants and repair shops
- Equipment operators, machinists, supervisors, and managers
- Engineering colleagues
- Machinists and plant managers in repair shops
- Managers at plants I've performed work for
- Technical experts in companies I've worked for
- Technical experts in outside companies I've worked with
- Technical society leaders and members
- Equipment vendors who helped solve problems
- Participants in seminars I've presented or attended.

The list certainly isn't complete, but the contacts have helped immensely in providing data and advice when solving difficult problems.

Throughout the book you will see how some of these contacts have been made. The value of these people can be realized when they leave or retire from a company. Calling and hearing they are no longer with the company was always a letdown for me, like losing a good friend. Sometimes, I could locate them again and the contact continued. When you cannot locate them, that particular level of expertise is now missing and can't be replaced nor is the friendship. Years of experience were needed for the advice the contact provided and it is doubtful others will have experienced and solved the same type problems.

As a consultant and as a company engineer almost all of my work has been obtained from networking. Friends or others I have done work for pass my name on or recommend me for jobs. Very little of my consulting effort was obtained by someone just seeing my name in print.

REFERENCES

1. Bloch, H.P., Geitner, F.K., Machinery Failure Analysis and Troubleshooting, Gulf Publishing Co., p. 343
2. Kepner, C.H., Tregoe, B.B., The Rational Manager: A Systematic Approach to Problem Solving and Decision Making, 2nd edition, 1976.
3. Bloch, K., Extreme Failure Analysis: Never Again a Repeat Failure, Hydrocarbon Processing Magazine, April 2009.
4. Wise, T., Tesla, Turner Publishing Inc., 1994.
5. Sofronas, T., Family and Friends in the Oil Patch, EP Press, 2004.

4

COMMUNICATING EFFECTIVELY

4.1 SPEAKING EFFECTIVELY AT MEETINGS

Being a good communicator is an excellent trait to have. Lawyers have it, politicians have it, and most business leaders have this ability. When an engineer can't make his or her point in a meeting, it's usually because they lack public speaking skills.

Being a confident speaker is also important, and I would recommend that everyone take a seminar in public speaking. The best advice I can give is to know your subject and audience well. Recording yourself as you rehearse will illustrate the necessity. Too many "um", "er", "uh" or talking too fast or slow, no enthusiasm, or being fidgety all will be quite evident as you watch yourself. Try using hand gestures to loosen yourself up. Correct these and you will be viewed much more favorably.

There are things you can do when you have to make a presentation. Most important is to prepare and to know your subject. You should have practiced what you are going to say and have note cards or a condensed computer presentation. Such a presentation should have the key points or photos to spark your discussion. Reading off of the screen word for word is highly discouraged. It's just plain boring to the audience. In addition, it's discourteous to them. After all they know they can read. Your presentation should supplement what's on the slide in your own words and style. Try being as relaxed as you can and have a friendly disposition.

The next very important task is to be able to anticipate and address any questions that the audience may have. For important meetings, make a list of 50 difficult questions that might be asked and develop a useful answer to them.

Survival Techniques for the Practicing Engineer, First Edition. Anthony Sofronas.
© 2016 John Wiley & Sons, Inc. Published 2016 by John Wiley & Sons, Inc.

For example, some questions that have been heard when presenting the results of a failure to top management are as follows:

- Have you contacted our other divisions to alert them of the problem?

    You can do this by issuance of your findings or in divisional meetings. Make sure it's acceptable with your legal department.

- Have you called in an expert?

    This is a difficult question to answer especially if you're considered the corporate expert on the subject. Contact an outside expert or the equipment builder so you have an answer for this question. Vice Presidents might not know who you are or your abilities.

- When can we safely start backup?

    Now this is a question you don't want to be responsible for. This is a good time to mention what the team thinks when the time for a start-up is. The risk to you is too great if you provide a date and can't meet it or if it doesn't start up safely.

- Where else in our plants could a failure like this occur?

    This is a good question but could be difficult to answer depending on the number of plants you have. For example, if it were a pump fire one plant might have 1,000 pumps and the company might have 20 plants.

- Has this happened to anyone else in industry?

    Probably so, but who? You can ask around, but others don't volunteer information on their sites' failures because of liability issues and pride. Bringing in an outside consultant with similar experiences might be a good way to get additional data.

- How did we ever let this happen?

    Tread carefully answering this question and think it out well. I'm sure you didn't cause the failure so don't try to answer this one or it may become yours. Causes can be anything from lack of maintenance or someone making a decision who didn't know the technical details. There is no good answer and this might be a good time to discuss the team findings not your opinion.

You must anticipate and answer such questions in a politically correct and respectful manner. This might not be the time for humor.

It's important to listen carefully to what the questioner is asking you. Don't be afraid to take notes to be sure you understand. All too often we are so anxious to answer the question that we don't hear it well or hear it wrong. It's not good to hear "That wasn't my question!" from your management.

Notice there are not many questions on the details of your troubleshooting procedure or the analysis. This is probably what an engineer would be prepared to talk about. Know your audience. A technical group would be asked different questions than a management team.

When it's necessary to call on a colleague who is in the meeting to help answer a question in an area he or she participated, let them know you will do this before the meeting. This is especially true at high-level meetings where careers could be affected. While you have prepared your answers to questions well, your colleague could be caught unprepared, which would be most unfortunate and embarrassing.

Again, it's important to take notes and say, "I don't have that information right now, but I will get you an answer," when you can't answer an unexpected question on the spot. You will be respected for that.

As was said previously, always see how the experienced senior personnel present their work and answer the questions that are asked. Attend technical presentations at conferences if you are not a participant in meetings at your work place. Practice in presenting presentations and fielding difficult questions is the best way to learn this skill.

4.2   EFFECTIVE WRITING SKILLS

Also of extreme importance are your writing abilities. Writing courses were given in college, but companies have a special way they prefer reports written. It's usually nothing like the college courses. So read the technical reports others write in the company you are in and follow the lead of the good ones.

Final reports for the various companies and even for consulting reports usually have a section at the beginning called a Management Summary. This is a very concise section not longer than a page and presents clearly the problem that was to be addressed, the findings in nontechnical terms, one or two possible solutions, and the next steps if any.

The rest of the report would be in terms the team would be interested in and can be quite technical. Interim reports usually contain more technical detail as they are written for the technical team.

I've always been one who doesn't enjoy long speeches, tedious articles, and reports when a few words will suffice.

Many years ago, a manager taught me a method called "Precis Writing" or precision writing. At the time, it was a glorified viewgraph type presentation meaning there were no unneeded words. It is a high density presentation because if you missed one word you could miss the essence of the presentation since it's only a guide for you to elaborate on.

When I am writing a technical magazine article in a one-page format, it may start out as several pages. When it's done, it's one page with some graphics included.

The process is to review it several times striking out words, equations, and sentences that don't add meaning.

Minimizing the use of unnecessary technical terms also simplifies the message. Here are two examples that illustrate this. Consider the two possible sentences, "Now calculate the system natural frequencies" or "At this point of the analysis the engineer needs to utilize analytical techniques to calculate the system eigenvectors and eigenvalues using either the matrix method or the Holzer method." One contains 6 words and the other sentence contains 29 words. All that was desired was to know that the natural frequencies needed to be calculated.

There are times when detailed information as well as references are necessary like in a technical paper, this book, or a doctoral dissertation. The reader needs to be able to verify or elaborate on the works. This is not the time for precis writing.

There are limits to brevity. For example, if the information is reduced to such a simplistic form that it is of no practical use to anyone, then there is no need in presenting it. One becomes "A Pontificator of the Obvious."

The title of the article is also very important. As is sometimes said one of the best forms of precis writing is the headlines of a major newspaper. In just a few words the essence of what is in the article is disclosed. A technical article title might read as "Electrical Faults Can Cause Shaft and Gear Failures." This is a better title than "Shaft and Gear Failures." Yes, there are more words, but they are useful words as they clearly describe the contents.

4.3   LEARN TO LISTEN

Remembering what someone was saying had been one of my problems early in my career. Yes, I got the gist of it most of the time but missed quite a bit. This happened at work and at home. It depended on the situation and if it was something I was interested in I would try to listen intently, but if it wasn't my brain would start to wander. I would start thinking about other things like what was for lunch. As the speaker you can usually tell when this happens to a listener or your audience as their eyes seem to go blank and glaze over.

What happens is that as the person is speaking to you, you hear a key point and then you are focused on that point because you have thought up a reply. You're so intent on perfecting your reply to illustrate your importance that you hear no more.

Here's an example. A friend is talking about his vacation and all that he has seen. Mid-way through the conversation he mentions Paris, France. You stop listening after you hear this because you have been to Paris and had something funny happened to you there. No details on the remainder of his trip are heard because your mind is too busy plotting out how you will impress him with your wonderful story. The rest of the discussion of his trip is missed, which is unfortunate as there were probably some really interesting gems that you will never know.

In an engineering meeting or when listening to a senior person, this can be quite devastating to one's career. When someone is discussing a problem and you only hear what you want to you may get it all wrong.

Try and focus on what is being said and not let your mind wander. Take copious notes at the meetings as you listen. You have to learn how to take notes because you can become so engrossed in taking those notes that you forget what was being said. My note taking is a sort of shorthand of important facts. As a private pilot I have to read back instructions given to me from the tower while flying to the airport. It's a busy time and you don't want to get distracted. When coming in to land I may hear something like "Contact tower on 118.35 when on a 3 mile right base for runway 14 and iden 1200." I might jot this down on my pad as 835twr3rb14 id12. I understand my shorthand and have only 15 characters to write down instead of 72.

After writing and reading about how to listen, I must say I still have problems with this. The only thing that seems to help me, if I can't write it down, is to try to remember at least two important points that were said during the conversation. This helps some, but I sure would have loved to have had total recall during my life. I'd probably have a lot more friends.

5

PROBLEM SOLVING AND DECISION MAKING

One of my clearest memories of a problem solver is when I first started working for a company that came up with new products for automotive use. The company had an older gentleman who would sometimes be sitting under a tree outside my window eating an apple. He would be looking at the sky, birds, and nature in general at various times of the day. I asked who he was and it seems he was a fellow with many patents for the company. He would go into his manager's office periodically and have a napkin or piece of paper with an invention of some sort on it, give it to the manager and say, "Try this idea I think it will work." It usually did and he had made the company quite a bit of money. I'm sure there is something he invented on the car you drive today.

We all may not have this inventive skill or problem-solving ability but fortunately in engineering there are other ways to solve problems and make decisions and some will be discussed here.

5.1 WHY IS THIS SECTION IMPORTANT?

In engineering, problem solving is one of the most important activities an engineer will perform. It is a process that involves discovering, analyzing, and solving problems with the goal of finding and implementing a solution that best solves the issue. While having all the skills needed to try and solve a problem is wonderful, unless you are methodical you may have many erroneous and costly attempts to determine the cause.

Survival Techniques for the Practicing Engineer, First Edition. Anthony Sofronas.
© 2016 John Wiley & Sons, Inc. Published 2016 by John Wiley & Sons, Inc.

I enjoy solving mysteries in my mind and especially enjoy Sir Arthur Conan Doyle's creation, the consulting detective Sherlock Holmes. Sherlock uses logical reasoning and science to solve his cases. Holmes states, "When you have eliminated the impossible, whatever remains, however improbable, must be the truth." Well, we needn't go that far to determine why something failed, but our methods won't be much different.

Problem solvers, decision makers, and good communicators are highly sought after in engineering and hence given its own chapter here. When equipment fails, it stops production or can cause injury and someone needs to come to the rescue. Without this ability, my career would have been severely limited. We discussed how a good analysis and logical, methodical thinking are the tools we need. Happily these tools can be learned and work even when we are new to engineering and have little experience. Watch successful problem solvers and decision makers in your organization and see how they do their job.

5.2   THE SIMPLEST SOLUTION FIRST

Many of us have watched the repair of our automobiles. They are now so complex that they need to be analyzed with a computer system. Every year some new electronic controlling device has been added. A colleague of mine had a new truck with so many new electronics that one signal didn't know what the other was doing and would malfunction. The poor folks in the automotive shop had no idea what to do or how to troubleshoot the problem and neither did the manufacturer. The fix was to replace just about everything and it still doesn't work quite right. His fix is going to be to sell the truck.

Replacing parts is one troubleshooting method, usually replacing the least costly first. Finding others who have had similar problems and have solved them is another route to take. In engineering, that would be similar to bringing in a consultant to solve your problems. It works but doesn't do much for your career and certainly isn't gratifying. Bringing in consultants shouldn't be what engineers are hired to do.

My computer wouldn't work meaning the screen was dark. I watched as the computer technician diagnosed the problem. The first thing he did was look to see if the plug was in for the screen. The next thing he did was turn it on and off several times rebooting the system. Then he checked to see that the cables were connected properly. Finally, he got down to his methodical troubleshooting procedure first asking me a lot of questions. You see he checked the simplest things first.

Recently, I was about to plant some flowers and took out my gasoline-powered cultivator to churn up the ground. It hadn't been used for several months and of course it wouldn't start. A call to a small engine repair shop indicated troubleshooting and repairing would cost about $150 plus parts. A new cultivator was $250.

Well that didn't seem reasonable to me since they didn't even know what was wrong, so simple things first. Was there spark, air, and fuel? Fuel and air were present and a new spark plug was being used. There was no spark at all even with the ground wire disconnected. All that was left was the solid state ignition module since there were no longer points and a coil to worry about. The cost of a new module was $40, which seemed reasonable. So for 1 h of troubleshooting and the cost of the module I was back to planting my flowers. Not quite a major engineering accomplishment, but it does show the process of simple things first and part change out when all else fails. In this case, it was cost-effective, fast and I was back in business again.

An experienced problem solver will usually be very methodical and resourceful. My grandfather was a farmer and he could always figure out how to make his broken farm equipment work. He could also modify his machinery, so it would work better and be more productive and he didn't even go to high school. I've noticed that creative problem solving usually is born out of necessity not education. For my grandfather if he didn't fix his equipment, his cows wouldn't have hay, his corn and tomato crops would fail, and his family would go hungry.

Now unlike my grandfather if an engineer doesn't solve a problem, it may not cause his or her family to go hungry. However, if the company is losing vast amounts of money or someone is injured, it could be perilous to their present employment. So there has to be a better way that an engineer can depend on to solve problems. Fortunately there is and here are some methods.

5.3   THE 80–20 RELATIONSHIP

Before getting into a suggested problem-solving method, I'd like to mention a rather old concept called Pareto's law after economist Vilfredo Pareto who wrote a paper about it in 1896. This relationship is sometimes called a law, rule, distribution, or principle and states that about 80% of the total effect will come from only 20% of the members in that group.

For example,

- 80% of failures are due to 20% of the causes.
- 80% of the injuries are due to 20% of the hazards.
- 80% of sales are due to 20% of sales staff.

Management sometimes uses this relationship to determine what areas to put the most effort into [1]. For example, if you're developing a technical library and have recorded the most used references, it might not be cost-effective to have your resources work on the last 20%. The obvious problem is determining when you have gotten the 80% most important.

This 80–20 relationship appears with surprising regularity and is worthwhile to remember when troubleshooting, problem solving, or making decisions even though the ratio may not be 80–20.

Consider, for example, centrifugal pump failures. Major causes that can occur are as follows:

1. Cavitation-related problems
2. Lack of lubrication

3. Poor repairs
4. Wrong process lineup
5. Alignment
6. Defective system or pump design
7. Coupling failures
8. Bad bearings/inadequate lubrication
9. Seal failure
10. Not following operating guidelines.

Historical data at one plant, when reviewing the history of 200 pump failures indicated that 2, 8, and 9 were the most experienced. Not quite 80–20 but close enough to illustrate a point (70–30). The point is to look for the primary causes. Depending on what power microscope you look at a problem with the more possibilities you will come up with. Many of these will be trivial.

Table 5.1 represents some of the more memorable failures the author has analyzed for the petrochemical and transportation industries. Each item consists of approximately 30 failures and certainly does not represent any statistically sound sampling.

TABLE 5.1   Failures and Pareto Distribution

Item	Cause (%)	Cause
Gasket 80% failure due to 40% of causes	50	Too low bolt load
	30	Improper assembly
	10	Wrong type
	5	Too high bolt load
	5	Damaged
Gears 55% of failures due to 33% of causes	33	Overloading
	22	Lubrication
	15	Design
	10	Misalignment
	10	Heat treatment
	10	Poor quality control
Ball and roller bearings 75% failures due to 60% of causes	35	Lack lubrication
	20	Improper assembly
	20	Excessive loading
	20	Defective bearing
	5	Other
Bolts and studs 85% failures due to 60% causes	35	Low preload
	30	Corrosion
	20	Overload
	10	Wrong design
	5	Poor quality control

TABLE 5.2   Subjects in Site Statistics

Analytical Modeling	Reciprocating Compressors	Gaskets	Pumps
Bearings	Equipment repair	Life extension	Vibration
Bolting	Extruder equipment	Piping	Seals
Centrifugal compressors	FEA	Pressure vessels	Wear of equipment

It does reenforce that many times problems or failures are caused by only a few of the variables.

So from this data the wide spread from the 80–20 relationship in engineering failures can be observed. The relationship is useful to realize that at some point obtaining data will have a decreasing benefit and also observing what data is useful. It's a good way to document your failure data to see if it follows this type distribution. For example, if lack of lubrication at a remote site causes most of the failures maybe the engineer should look into oil mist lubrication if applicable.

As a last example of this relationship I'd like to review several years of interest in my website and the site statistics. On the website, I have the following subjects shown in Table 5.2, which I can observe what readers were interested in.

Over a 4-year period, there was 80% interest in 45% of the subjects. The subjects were piping, vibration, reciprocating compressors, pumps, pressure vessels, bearings, and analytical modeling. All this really means is that the readers of the site referenced these topics most. This indicated to me that attendees of seminars I presented would also like to hear about these topics since they were probably seeing failures in these areas.

5.4   THE FIVE WHY's USED IN PROBLEM SOLVING

I've used this approach in gathering data and its intention is to get to the root of a problem. Credit is usually given to Sakichi Toyoda and was originally used within the Toyota Corporation. In its original form, it probably was started by the classical Greek philosopher Socrates (470–399 BC) as a way to stimulate critical thinking.

The method has the questioner asking the one question "Why" after each of five answers.

For example, let's say a bearing failed and you're talking with an operator at the unit. After asking what happened and her reply was that the pump had failed you ask "why" and the answer is the bearing got hot. You ask "why" or how come again and the answer is it ran out of oil. You ask "why" again and the answer is because it's a remote pump and no one made the rounds out there. You ask "why" and the answer is because it was raining all week. Only four whys but probably enough to look into some remote lubrication system such as oil mist or better preventative maintenance.

So why do they call it the five why approach. Probably because that's how long it takes to get some kind of useful answer. Mine never really got past four why's because the person I was questioning usually got annoyed.

Actually I like the method because I have Greek ancestry and have read a little on Socrates. Unfortunately for him because of his tormenting way of questioning, some accounts say he was eventually forced to take hemlock poison by those he tormented. This is probably a good reason to limiting the number of "why's" you ask.

In retrospect, I recall this method being used by our children and grandchildren. When they were 5 or 6 years old this line of "why" questioning would go on and on. It usually ended when we ran out of answers, they were satisfied with an answer or they lost interest. It was their way of learning and understanding too.

5.5 BEING THE DEVIL'S ADVOCATE

Another method to aid in problem solving is the role of being a Devil's Advocate. It's a religious term [2], so I don't consider it improper to use the devil's name. My interpretation of this in problem solving simply means looking at something another way to cause a debate on the subject.

For example, I was once analyzing an engineering failure and concluded what was the most probable cause of the failure. My colleague Dick, being the Devil's Advocate said "What would happen if you had received the wrong data from the client company?" He was really performing a "what-if" type analysis. Someone just turns the situation around to make you think of other possible causes. It's good to have a trusted colleague look over safety-incident-related failures like this.

5.6 AN ENGINEERING APPROACH: USE THE SCIENTIFIC METHOD FOR PROBLEM SOLVING

The best strategy for solving a problem depends largely on the situation but in engineering since the problem might be why some equipment has failed the Scientific Method is typically used. This is appropriate since this is the problem-solving method most scientists, engineers, and researchers of all disciplines use in one form or another. You probably first used it in a science experiment. It's the method of using observation and experimentation and has been around for a long time. Before its use, problems were solved by guessing and nothing could really be proven. Rather than try to describe the procedure in words, I'll use an actual case history to explain the method.

Before you start to solve a problem it is a good idea to make sure there really is one! Has something really changed or has a baseline never been established.

This isn't an empty statement. Many pumps have been rebuilt at considerable expense due to a bad $20 pressure gauge. Likewise, much time and money has been expended on items such as unacceptable leakage rates when no basis on what historically was the rate had been established.

Let's say a problem develops. Here's a method that has been useful in several variations:

1. Define the problem.
2. Gather data.
3. Organize and analyze the data.
4. Make an educated guess based on the data.
5. Verify your guess by tests, additional data, and calculations.
6. Implement a solution.

Actually this is just a form of the Scientific Method (substitute hypothesis for guess).

Let's look at 1–6 in some detail and then use the method in a case history.

1. *Defining the Problem*: Be careful here as many of your sources will start defining the cause and the solution. This is okay if the problem is identical to one that has occurred and all sources agree. Unfortunately, few problems are like this.
2. *Gathering Data*:
   - Gather facts and history from all sources.
   - Don't assume anything.
   - Screen the facts based on the reliability of the sources.
   - Based on how the process, equipment, similar equipment, and so on usually runs determine what has changed or is different. Ask yourself if anything has recently changed meaning has there been a deviation.
3. *Organize and Analyze the Data*: When the problem solvers are somewhat familiar with the equipment or process, the following table has been helpful. It's usually helpful to add any quick tests or calculation necessary to answer what the significance might be.

Observation	Significance
1. -------	● -------
2. -------	● -------

4. *Educated Guess*: Based on an intelligent review and screening of the data, this is where you make your best guess on what you think the cause is. Remember, it should be supported by data.
5. *Verifying Your Best Guess*: Here is where you try to prove your guesses wrong one by one. It's a good place to utilize some of the engineering calculations if applicable. One example might be that a part failed due to a cracked shaft.

    An overload and fatigue calculation could verify or dismiss this guess.
6. *Implement a Solution*: Solving a problem is not of much use if you don't implement a solution, so it's not a problem anymore. Another way of saying this is once the cause is identified, it must be removed.

    The best way to describe the method is with an actual case. The six previously mentioned steps will be followed.

This example relates to a centrifugal pump, which is a pump widely used in industry and kept tripping out on overload.

It was critical to production in that it limited production when it was down. Unexpected tripping also caused the system to plug, which resulted in an unscheduled downtime to clean. This is a very undesirable situation and one that causes the wrath of management to appear. Figure 5.1 illustrates a typical plugged pump, which is much smaller than the one that failed but whose plugging mode was the same. The impeller could not be turned at all.

1. *Problem*: One of the four slurry pumps trip out on overload during operation.
2. *Data (from Various Sources, Unscreened)*:
    - Same problem as in past, it's the slurry.
    - Motor is bad, pull it out.
    - Pump is bad, pull it out.
    - Operators lining it up wrong.
    - There is no problem.
    - It's the process.

    A lot of other information not particularly useful was obtained.
3. *Organize And Analyze*:

Observation	Significance
- New pump and motor installed 2 days earlier to try to solve problem.	- Unlikely same problem again on pump and motor.
- Not happening on other pumps.	- Same pump type but different slurry and controls.
- Problems in past never really just one cause. Has been due to slurry concentration, bad motor, bad pump, air leaks, and so on.	- Going to need tests to isolate this one. This is why there are so many opinions because been multiple problem causes in past.
- Ran a quick pump curve. Found 100 A draw (12% above full load) at design point, on water, no slurry.	- Drawing high amps not a slurry problem.
- Uncoupled motor, no load, 20 A, coasts down okay, pump turns easily by hand.	- Motor okay, pump packing not too tight, pump okay.
- Coupled motor and pump. Ran a pump curve of amps/flow. Amps high for flow.	- Amps showed too high for indicated flow. Since motor and pump okay, might be instrumentation.

4. *Educated Guess*: Since the problem solvers were well versed in the system the following was stated: "The flow is really higher than we are reading and since the motor and pump appear okay, it is the instrumentation." A higher flow explains the high amps.
5. *Verifying*: A test was run utilizing an alternate method for measuring flow. The amps were correct for the calculated flow. The instrument was bad.

6. *Implement Solution*: The instrument was recalibrated and put on a quarterly calibration schedule as were the other similar pump systems.

Note that the key points to this actual problem solving was organization of data and verification of most probable cause (guess).

Also note that a pump and motor had been rebuilt and replaced unnecessarily. The cost was approximately $10,000. The problem solving took 6 h.

This procedure can be used on a large number of problems both scientific and also managerial. It is exceptionally useful when others are inputting data into Steps 2 and 3.

One very important element of the method is that it produces a well-thought-out, logical approach that can be documented and reviewed by others and input from all points in an organization can be included.

Areas that aren't well investigated in this type analysis are organization-related problems that require a different approach in addition to the one discussed here.

Very rarely is a failure due to only one cause. Several interrelated factors usually combine in a chain of events that result in a failure; however, there is usually one major contributor.

Figure 5.1   A plugged centrifugal pump.

5.7   YOU NEED TO KNOW THE WHOLE STORY

My experiences with consulting for various companies is why this section is being written. When you are part of a company, you usually have time to get to the scene of the failure and collect data and interview personnel that you know. Even at meetings, you know the capabilities of people and you may be the one to select those on the investigation team. Noncritical failures for which most investigations were are the type where production has been limited. Fortunately, there have not been any accidents, fires, or liabilities of any kind. Management just wants the equipment on line quickly with no chance of a repeat failure. When I was a permanent employee of a company, the failure analysis procedure usually went something like shown in Figure 5.2.

As a consultant, you were usually called in later as shown. This is difficult because you don't have all the history of the machine and haven't seen the failure. There may not be any failed parts available, parts may never have been to a metallurgical examination, the computer data may not exist, and there may not be photographs of the failure region. The machine may even be back operating. In other words, you don't know the whole story. You will have to understand that your analysis may not be complete as the history of the machine before the failure occurred is not known. This is discussed further in Chapter 7.

Once as a consultant I was asked to investigate a noise in a gearbox that kept getting louder with an increasing vibration. It had been rebuilt several days earlier by

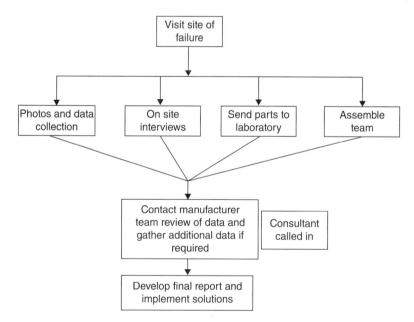

Figure 5.2   General failure investigation flow chart.

Figure 5.3   Wreck of a gas-engine compressor.

an outside repair shop. On disassembly I found the culprit. A wrench was found all mangled and had broken out some teeth on the gear. No problem finding out how the wrench had gotten there. On the wrench was etched "Stolen from B.W." which were the machinists initials. Unfortunately, this wasn't the cause of the problem before the machine was repaired. The true cause was a bad coupling that failed the bearings. No failure analysis had been performed at that time.

While the detailed problem-solving approach is useful sometimes just following the procedure in Figure 5.2 works. Figure 5.3 is a wreck of a component called the "doghouse" of a 3,000-horsepower gas engine compressor that had just occurred when called to the scene. It's called the "doghouse" because it looks like one. It contains components to convert the rotating motion of the engine crankshaft to the linear motion need to power the compressor piston. The debris inside is the remains of a linear bearing pad called the slipper. I was told there wasn't time for a detailed analysis as they had the new assemblies ready to go in as production had been greatly limited.

This tends to violate a rule that is "You Don't Have Time To Hurry Up!" This simply means if you do things in a rush you will probably miss something and have to do it again thus wasting time. The mood of management was such that it was prudent not to repeat the rule and got to work just using the procedure shown in Figure 5.2. In summary, the operators said they heard a loud clicking sound and then a loud "boom" and clunking. A quick review of the machine's 30-year history had no other failures of this type. The engine was still running when they shut it down as the compressor piston rod had broken and left the rest free to reciprocate back and forth (the clunking). The sliding bearing called a slipper was bolted to the bottom and the holes had thread imprints in them and the bolts had sheared. A check on the bolts on the other three "doghouse" slippers indicated that the other 12 bolts had not failed but 2 bolts were only finger tight. It was concluded that the bolts had loosened, sheared due to the impacting (clicking sound), and the slipper came loose and shattered (the boom) since it was cast iron. The manufacturer's representative was contacted and thought this could be possible. This usually means they have seen it happen before but don't want to admit it for legal reasons. The bolts were all torqued and safety wired, and a routine check was made at every downtime until it was decided they were remaining tight.

This was not a very detailed analysis, but it seemed to solve the problem. A check with other owners of the same make machine said they too had cases where the bolts would come loose but had no slipper failures. It was decided to dowel the slippers as well as bolt them at a later scheduled downtime.

5.8   FAILURE ANALYSIS AND ACCIDENT INVESTIGATIONS DIFFER

Most of the analysis shown in this section has been used to augment equipment failure analysis sessions. My specialty is in an analysis of the equipment for causes. As an employee and then as a consultant, I am usually part of a team composed of plant personnel. The purpose of the team was to understand the failure causes and return it to service as quickly and safely as possible. As mentioned, the failures were usually only limiting production.

Accident investigations are quite different and are used to thoroughly and exhaustively investigate the causes and the contributing causes. Unit explosions, fires, fatalities, all would come under these type investigations. Contributing causes are the factors that lead to the failure and involve multiple interrelated factors including management system failures. These type reports can take months or even years to complete.

A highly structured approach is necessary and team selection is an important part. Examples of this are shown in Refs [3–5]. The reason for this process is that the problem may have been due to the engineering and the organizational structure itself. The complexity of a national tragedy investigation is evident by reviewing Ref. [3], which shows all the mechanical, organizational, and governmental decisions that interacted to cause a catastrophic event.

Committees and teams each handling specific disciplines are well planned out with their roles documented. For example, in the case of the National Transportation Safety Board (NTSB) in air crash investigations there's a standard procedure, which involves the Investigator-in-Charge and Specialists responsible for clearly defined portions of the accident investigation. The teams are called Go-Teams and are always ready for action. The following are examples of their duties after the wreckage is recovered but not of the whole process:

- *Operations*: The history of the accident flight and crewmembers' duties
- *Structures*: Documentation of the airframe wreckage and the accident scene
- *Power plants*: Examination of the engines and accessories
- *Systems*: Study of components of the plane's hydraulic, electrical, pneumatic, and associated systems, together with instruments and elements of the flight control system
- *Air Traffic Control*: Reconstruction of the air traffic services given the plane, radar data, and transcripts of controller-pilot radio transmissions
- *Weather*: Gathering of all pertinent weather data for a broad area around the accident scene
- *Human Performance*: Study of crew performance and all before-the-accident factors that might be involved in human error, including fatigue, medication, alcohol, drugs, medical histories, training, workload, equipment design, and work environment
- *Survival Factors*: Documentation of impact forces and injuries, evacuation, community emergency planning and all crash-fire-rescue efforts.

Most accident investigations, such as the NTSB, follow various forms of the scientific method. There is not much detail in accident investigation methods on how the specialists are to do their job, only on what to look at. That is why the investigators must be proficient in their respective areas.

Unfortunately, no matter how thorough an investigation is, it can still be corrupted by the politics of the situation. This has been evident when reviewing accident reports. Sometimes, items that need to be addressed are left out of the final report since they might cause an embarrassment to the organization or loss of future funding. When too many costly items or changes are recommended to the present system, only the least costly or least politically sensitive might be implemented. Changing a safety item to a maintenance item, to expedite a schedule has occurred and influenced the chain of events that resulted in the accident.

There is not much that can be done about these type problems other than to be aware they occur and address them as best as the committee, team, or individual can.

When reviewing historical accident reports, I always looked for key lessons learned to see if they have been addressed adequately and implemented. Many had been addressed, but it was usually the implementation that was lacking. In some cases, the situation was ready for a similar event to occur again.

The intent of this section is to alert the reader that accident investigations require a specialized team that can undergo the scrutiny of a legal investigation. There are times when we should understand our limits as investigators and the excessive risk we may be undertaking.

5.9 WHY DECISION MAKING IS IMPORTANT IN ENGINEERING

Making technical decisions is one of the most difficult tasks an engineer has to do. This is primarily because we are always in fear of being wrong. As was mentioned, engineering is not without risks and if you don't take risks you will be stagnant in your career. Informed decisions are easier when you have all the facts required to make the decision. Unfortunately, we seldom do and many times the data we have is in error.

This is what is elegant about engineering. We have seen a method for problem solving and it is quite straightforward and can be tested for faulty data before making a decision.

There is also a method to allow you to choose, meaning make a decision on one out of many items you are comparing.

In problem solving, the decision came out of the problem-solving process. When there are several possible choices and your job is to pick the most appropriate choice, a systematic approach is required for all but the simplest cases.

5.10   DECISION ON SEVERAL CHOICES

In this section, a straightforward method is presented that weights and scores factors important to the decision. Since several key people can be solicited for their inputs, it then becomes a team effort decision.

Rather than going into the details of scoring models, which have been around for a long time, let's work an actual problem and it will become self-explanatory.

In this example, we need to make a decision on a piece of machinery to replace equipment with a poor service history. Six pieces of lifting equipment were compared, which were somewhat like an elevator lifting a barrel and dumping it into a mixing tank. Any unit had to cost less than $20,000 including installation. For simplicity, this example only shows two of the machines that were evaluated.

Looking at Table 5.3 the factors being rated were agreed on by a team who would be using the equipment as were their importance. A 10 weight means it's very important.

After the factors and weights are established, the individual making the decision can review the equipment information and give it a rank. If it was excellent on satisfying the factor being considered, a 10 rating would be appropriate. Likewise, if it poorly satisfied the factor 1 might be a good choice.

The final score is simply the sum of the weight times the rating. To get this into a percent of the ideal or best that can be obtained, divide the score by the sum of each maximum weight times maximum rating and multiply by 100.

$$\text{Ideal} = 10 \times 10 + 10 \times 10 + 8 \times 10 + 5 \times 10 + 10 \times 10 + 5 \times 10 = 610$$

Cable design $1 = (239/610) \times 100 = 39\%$

Hydraulic lift $2 = (445/610) \times 100 = 73\%$

The other four designs done in a similar manner scored 63%, 61%, 63%, and 58%. The other designs were fair in that they were much better than the original cable design (39%) but poor compared to the one that scored 73%.

TABLE 5.3   Decision-Making Ranking Table

Factor Rated	Weight 1–10	Present Cable Design	Rating 1–10	Hydraulic Lift	Rating 1–10
Operating safety	10	Binds, cable breaks	4	Restrictor for easy let down	8
Mechanical/operating reliability	10	Above has caused poor reliability	3	Should be good simplicity	7
Easy to use	8	–	6	–	9
Low maintenance cost	8	Has been a service factor problem	2	–	6
Design simplicity	5	–	4	Hydraulic cylinder and chain	4
Minimum space	5	–	7	–	7
Proven design	10	New units have cable problems in 1–2 years	2	Much like fork lift	9
Good housekeeping	5	–	6	–	6
ΣWeight × Rating	–	–	239	–	445
% of Ideal		(239/610) × 100 = 39%		(445/610) × 100 = 73%	

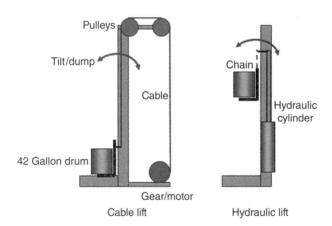

Figure 5.4   Barrel dumper selection.

While your engineering judgment will control based on past experience with similar type equipment, such a systematic organization allows you to screen the choices.

As a point of interest, three hydraulic lifts of the type that scored (73%) were purchased and have worked flawlessly with no safety incidents in 4 years. The best compliment has been no complaints from operators or machinists. Figures 5.4 and 5.5 show the original design and selected design, respectively.

Figure 5.5   Installed selected barrel dumpers.

The scoring model method has its problems since the weighting and ratings are susceptible to personal preferences. This must be recognized when using this approach.

5.11   THE IMPORTANCE OF PERSONAL CHECKLISTS

When I took up flying, I realized my memory alone was not going to be good enough. It was self-preserving to see if the engine was developing full power and the flight controls were functioning properly before leaving the ground. So before departing I run through my written checklist of 15 items mentioning each audibly while touching each gauge or knob, much to my passenger's amusement. Becoming complacent about checklists can be dangerous.

Checklists are just as important during equipment start-ups. Whether you are starting up a 20,000-horsepower turbine or a small pump that is handling a hazardous product, your own personal checklist is a valuable tool.

Major projects have well-thought-out checklists called "punch lists." I'm not talking about these here. I'm talking about your personal ones that help you from making those career limiting mistakes or as sometimes called CLMs. A commercial pilot who forgets to review his checklist and doesn't put down the landing gear probably will survive but his career might not.

Here are two condensed checklists I've developed and used over the years. Like everything else they were based on problems seen on systems I had been involved with or inputs that I have gotten from others.

Starting Up New Piping Systems

1. Is there a critical flange list and has it been followed?
2. Are all supports in contact with piping and shipping stops removed from bellows?

3. Do the welds look OK? A poor looking weld usually will fail especially socket welds.

4. Are hot bolting procedures available in the event of a gasket leak?

5. Do the piping stress isometrics make sense with what has been built? Are things where they are supposed to be?

6. Is everyone positioned safely during start-up? There is no need to have someone working on piping that will become very hot.

7. Have you "walked the line" before start-up, scanning for items that just don't look right? For example, a pipe that goes nowhere or blinds still in place. It happens!

8. Has a hydro-test been done and have you reviewed the results on what was found and was it corrected?

9. Remember if any of the piping is high pressure steam, any steam leaks could be invisible and lethal. Have a way to verify there are no leaks and make sure everyone is aware of the danger involved.

10. After start-up have you "walked the line" again checking for leaks and excessive vibration of the piping?

11. Is the piping clean inside, meaning have blow-down tests been performed?

Compressor and Pump Start-Ups

1. Have you read the manufacturer's start-up procedures on the motor, compressor or pump, and gear unit?

2. Has the alignment been checked and do the numbers make sense?

3. Has the inlet piping been cleaned?

4. Is the lube system OK and functioning? Check the schematics.

5. Is the motor, compressor, and gear unit oil drain back to the reservoir correctly?

6. Are the seals piped up properly with the correct product used?

7. Is everyone positioned safely, meaning out of the way of leaking product, flying couplings or parts?

8. Has the vibration, temperature, and pressure instrumentation been checked out and limits established such as when should you shut down?

9. Has the motor been checked uncoupled by electricians for correct rotation and no load amps?

10. Steam turbine drivers should have their own checklist, but it's essential that the overspeed has been checked and there is a method to check for those invisible high pressure leaks that are lethal.

11. Have the designated spare parts been checked and are they readily available? A spare rotor that doesn't fit or has a gravity sag isn't of much use. Neither are corroded, damaged, or the wrong bearings or seals. Just because a vendor says it's on the shelf doesn't make it so.

While the list of items are small they require a considerable amount of effort to ensure they are correct. These are usually items that have caused problems on similar equipment and an expensive shut down was required or there was a safety issue.

Speak up early if something looks wrong. When it cannot be repaired before start-up, consider contingency plans. For example, if it's discovered late that there are hundreds of socket welds with poor penetration, consider a risk-based plan that will address the most serious faults first. For example, a high probability of a major pipe separation may need to be addressed before start-up while the possibility of a low risk small nontoxic leak might be flagged for inspection or repair at the next downtime.

Many people are involved in major start-ups and engineers should be aware that high-level careers are enhanced for on time, below budget start-ups. Those who cause delays are not popular. You should have the support of others and very good reasons and documentation for causing a delay. Severe compromises on safety are good reasons.

5.12   CONFIRMATIONAL BIAS OR SELF-FULFILLING PROPHECIES

This term came up in a book I was recently reading [6] and I had never realized how relevant it was in engineering. Basically, it's when you are looking up or testing data and favor your beliefs or what you want to select over what is correct. It has other more colorful names such as "cooking the books," or "fudging the data." A "fudge factor" is not the same thing. This type factor is included on purpose and is used to make the analysis closer to the experimental data. We need to be aware that confirmational bias exists in our choices.

Throughout this book and during our lives, there are many opportunities for confirmatory bias. For example, problem solving and decision making as discussed in this chapter require an unbiased attitude when reviewing the data. You can see how detrimental this might be when testing a hypothesis against the data as to why a failure has occurred. Remember you never prove a hypothesis correct, you just test it and after enough positive tests, you feel comfortable with it for practical use. However, just one valid experiment can prove it wrong. Drop an apple 1 million times on earth under normal conditions and if one of those times the apple flies upward you may have to rethink the hypothesis on gravity. However, for practical purposes, Newton's laws still would be useful.

Let's say you have done a detailed analysis and come up with an answer. You look through the literature and you unknowingly only choose the data that supports your analysis. You have therefore produced an erroneous analysis. Having had someone knowledgeable review your work would have been one way to help circumvent this bias.

This is also true for problem solving and decision making. The more people involved in the method the less chance for this bias to occur. Having a team select the weighting and ranking can help reduce the possibility that you unknowingly have "adjusted" the data to favor what you want.

I'm going to have several unbiased souls review this book before it's published. They will be people who know me and what I have done. It's important since this isn't a book of fiction and I don't want to unwittingly say things that are untrue. It's one way to keep me honest to my readers.

REFERENCES

1. Bloch, H.P., How One Manager Views Pareto's 80/20 Rule, Hydrocarbon Processing Magazine, September 2015.
2. Helterbran, Valeri R., Exploring Idioms, Maupin House Publishing, 2008.
3. Columbia Accident Investigation Board, Report Volume 1, National Aeronautics and Space Administration, Government Printing Office; August 2003. 248 pp.
4. National Transportation Safety Board, Aviation Investigation Manual, Major Team Investigations, November 2002.
5. Paradise, M., Unger, L., Taproot: The System for Root Cause Analysis, Problem Investigation & Proactive Improvement, 2008.
6. Eberhart, D.F., The Switch, Greenleaf, 2015.

6

HOW AN ENGINEERING CONSULTANT CAN HELP YOUR COMPANY

6.1   WHY USE A CONSULTANT?

With company engineering resources heavily used, the time required to analyze an unexpected failure can be demanding.

Nontraditional designs cause unique problems for a company's engineers because they may never have seen these type failures before.

The usual method to try to address the problem is to contact the equipment manufacturer, tell them what has occurred, and hope they can provide a solution. Sometimes, this works as long as it isn't something that will be costly for the manufacturer to fix or could result in legal actions to them because of production losses.

Many times the response from the manufacturer is as follows:

- The correct operating procedures aren't being followed.
- The machine has been overloaded or abused.
- This type failure has not occurred before.
- You must be operating outside the design limits.
- All the applicable design codes have been adhered to.

These responses are not of much help in fixing the problem. In many cases, the manufacturer may be correct, but the equipment owner may not agree.

The owner may ask for the design calculations in the failure area to be provided to them. The response from the manufacturer may be that its proprietary information

Survival Techniques for the Practicing Engineer, First Edition. Anthony Sofronas.
© 2016 John Wiley & Sons, Inc. Published 2016 by John Wiley & Sons, Inc.

and can't be supplied. Sometimes for specific processes this is true, but most times they involve methods readily available to engineers.

The following are typical failures the author has worked. The equipment was unique with little design information available or provided:

- Vibrating conveyor structural failures
- Material recycling choppers structural failures
- Rotating mixer/dryer structural failure
- Gearbox cracking, movement, gear pitting, and breakage
- Extruder screw wear and breakage
- Auger screw conveyor shaft failures
- Agitator shaft seal failures
- Gas-engine compressor wrecks

Sometimes, similar failures have occurred elsewhere in the company with approximately the same life. In engineering failure analysis, there is no such thing as, "they all failed with about the same life so it must be a coincidence." In engineering, there is usually a reason.

On new equipment the problem can be that the owner hasn't written a detailed enough purchase specification, so the manufacturer provided a low-cost unit to meet the specifications. Unfortunately, for new projects, the purchasing of equipment usually goes to the lowest bidder. To remain competitive, the manufacturer may have to leave out the "extras" that result in a highly reliable design. The usual rule for contractors is to meet but not to exceed the clients' specifications.

On uniquely new designs of larger power and geometric size, the manufacturer's smaller and highly reliable unit may have been scaled up. Many things don't scale up well such as tolerances and welds.

6.2   WHAT A CONSULTANT CAN DO

The consultant may have worked on similar failures for other companies and therefore has a wider experience base than others may have. In my case, there are usually few failures that I haven't seen occur somewhere and may even have worked on them.

On unique equipment, a consultant skilled in simplifying complex machines may develop an analytical model. This allows all of the loads affecting the failure area to be analyzed and easily explained. Failures unlike a new design allow the investigator to know exactly where to analyze especially if there are weld cracks. This approach is also used on reviewing a new design to anticipate future problems and hopefully address them early.

One output from such models is the expected life in years due to the loads and stresses. This is shown in Chapters 9 and 10. Defining methods to reduce these loads and stresses will therefore increase the life. While life expectancy calculations are not an exact science, they will show the effect of increasing production rates or changing the processed materials composition on the equipment's life and the sensitivity of the other variables.

With this type data produced and shown to the manufacturer of the equipment, the manufacturer may be more accommodating in providing the help required.

Consultants are also available for training purposes. Early in my career, this is the way my supervisor brought me up to speed. For example, I might do an analysis on a vibration system and then a consultant would do a similar analysis and we would compare notes. I discuss this in Section 11.4 on torsional vibrations. It was expensive for my company but was all paid back with the learnings I had received and the problems I could solve. Consultants who were experts in areas we had equipment problems were also invited to present seminars to the engineers and technicians on specific types of equipment.

6.3   THE COST OF A CONSULTANT

Experienced consultants are not inexpensive. One has to recognize that the consult maintains the software and experience that the companies' engineers may not have. Finite element, fracture mechanics, structural analysis, vibration analysis programs, and laboratory testing equipment are a few items necessary to do their job.

The consultants billed work should be compared with the following:

- The production losses, legal liabilities, and the repair costs for a repeat failure.
- The use of the engineering staff's time needed to do the analysis.
- The possibility of sharing the billed work with partner sites with similar failures.

Having a consultant available is like having a partner on the owner's failure analysis team. The consultant can provide valuable data to the team for identifying, explaining, and addressing the failure causes.

7

CONSULTING ENGINEERING AS A CAREER

While my career after retirement was Consulting, my true desire was to own a company that manufactured products. I wanted to build things. Two products were developed that had limited success but from which I learned much. The first product was a motorized tow for pushing small 2,300-lb aircraft in and out of hangars. The second was an electronic vibration monitor for recording the vibration levels on aircraft engines and noting adverse vibration and what it might mean. Both were technical successes, which meant they worked well and those who purchased them liked them. They were both, however, marketing failures.

Selling a product is an expensive proposition and advertising is very expensive. In the world of aviation, people who have small aircraft usually have the person that fueled up their aircraft, help them push it into the hangar and pull it out, and so a large market became much smaller. As for the vibration monitor, most private pilots don't understand the usefulness of knowing the vibration levels and think that if something is vibrating too much they will feel it. This is not so because the vibration level on the engine is 20 times more than the pilot feels in the cabin because of the engine isolation mounts. The problem became one of educating the pilots on the need. This is also very expensive.

Thus, my venture into a manufacturing company didn't work out well, but it was fun. Fortunately, I did the design and manufacturing of the motorized tow myself and

Survival Techniques for the Practicing Engineer, First Edition. Anthony Sofronas.
© 2016 John Wiley & Sons, Inc. Published 2016 by John Wiley & Sons, Inc.

I located a company that agreed to engineer the vibration monitor at no cost to me for a return on the sale of each one. At the end of it all I came out about even on both.

My lesson to the reader is if you don't try something you will never know if you will be successful. The "rush" you get wondering how many you might sell is unexplainable. Just make sure that you evaluate the financial consequences before you start and perform a detailed research into the market.

7.1   CONSULTING AS A CAREER

Early in my career, a friend owned a structural consulting company and needed some help in analyzing a multistory building in Chicago. I was asked if I could help at nights and since he was the owner and the pay was good I took the job. My company didn't mind since there was no conflict of interest. It was an interesting assignment as I was to analyze the support design for attaching the many glass window panes to the framework. It was challenging since you didn't want the windows to pop out of their flexible seals and fall to the ground from the 20th story in high winds.

A couple of years later, I read and saw on national news about a building whose windows had popped out in a high wind and fallen to the ground shattering into many pieces. Luckily no one was hurt. As I cautiously called the consultant I had performed the work for, he answered the phone with "No Tony it was another city and another consulting firm." He knew exactly why I had called and I'm sure his heart stopped for a few seconds too when he heard of the incident. This is a major reason why I didn't go into consulting in my early years. I just didn't want that anxiety and responsibility. Later as I became more knowledgeable and had performed many analyses, I realized it was just my inexperience that made me so hesitant. By then I was quite content working for large companies and all of the advantages and flexibility they afforded me and consulting didn't interest me as much.

Many engineers who want to try running their own business and have become known as experts in their perspective areas consider entering into consulting.

The idea of being your own boss and selecting the job you want is a good reason. They watch consultants who work with their company and say, "I could do that!" They probably could do the work, but there's much more to consulting than that. I have been consulting for 15 years after I retired so again much of what you read here is from actual experiences.

Consulting is a difficult area to enter into unless you have contacts. Many decide to consult for the company that they have just retired from. This is the most productive way since you already know the company, its problems, and the people involved. One must be aware however that your role in the company, who you report to and how much your input is valued will most likely have changed since your retirement. You may now be reporting to someone you once supervised and may no longer have the same respect by management since the management staff may have changed and not know your work history with the company.

Other engineers prefer to expand their area of expertise by consulting to other companies not only the one they were employed by. This is what I did and requires considerable more effort to obtain work. For very serious problems, companies prefer to utilize large established consulting firms even though your work history and experience may be a better fit. Cold visits to companies, published books and articles, or advertising usually don't result in a flood of jobs. Work is usually obtained by being recommended by someone who is familiar with you. After the first job, you will have established contacts with that firm and they will call you again if the need occurs and they have been satisfied with your past performance.

7.2 COMPENSATION WILL PROBABLY BE LESS THAN YOU EXPECTED

Most experts go into retirement or consulting with adequate financial support and those who want to start in the business should be adequately funded or self-sufficient also. The reason for consulting is because the specialists want to transfer their expertise to others by seminars or solving various problems and enjoy the work they are doing. In addition, consultants are frequently needed by industry to supplement their resource requirements for new projects or unexpected problems. It is extremely rewarding to have your expertise and experience utilized by others.

Compared to your annual salary and benefits as a permanent employee of a company, the yearly effective salary of a consultant will realistically be much less and without company benefits. There are several reasons for this. The first is that time will be spent looking for jobs and this will be at your expense. When one obtains a job, not all of the expense can be charged to the client as it would then be uncompetitive. For every 1 day worked 3 extra days may be required to understand the problem, develop the analytical model, research the problem, or have initial meetings with the client. It would be inappropriate to bill these items. In addition, the cost of software, maintaining your expertise, along with your technical affiliations, travel and attendance, and medical provisions are now covered by you and not the company you work for. One painful adjustment was not flying First Class overseas anymore, to keep the billing expenses reasonable to the client.

One method after reviewing a job for a prospective client is to develop a Job Proposal. This would outline what you could do to address the problem, present the deliverables along with billing and timing information. The client can review this with the team and see if it is acceptable. A word of caution, the consultant should already know how to do what is in the Job Proposal. For example, don't say an analysis will be done if you have no idea how to do it. This might require some unbilled work as mentioned previously. Do a simple similar analysis first to see if it can be done before proposing it.

When work by others is required, such as materials analysis, or a large finite element analysis, this work can be billed separately from your work, as you may not have a good idea on the costs. Having the client set up a separate purchase order for this work is usually the safest approach. You can then include following the work in your proposal.

It has been reported that up to 10% of jobs a consulting firm takes on will not be paid. Fortunately, this has not been the case for me since I only deal with known reputable companies. It is usually prudent to only start a job after a purchase order has been received.

7.3   HOW MUCH SHOULD MY BILLING RATE BE?

Having been an expert in industry, you have undoubtedly had to use consultants and know what the fair and accepted rates of individual experts are. There is quite a difference between the billing rate of a large consulting company, who have to include their overhead costs, and the individual consultant. You should use these as your guide and bill accordingly.

7.4   THE JOB CONTRACT

Some surprises may occur to those new to engineering consulting when they receive the contract from a company for a particular job. There will be much paper work associated with this when larger firms have had their legal department draw up this contract. Everything from contractor compliance forms, safety records of your firm along with Professional Errors and Omissions Insurance that you are required to have may be included. This is expensive insurance and is not a problem if you are a large consulting firm with many people working in the field. It is quite different if you are a one-person organization working from your office and possibly the client's office and never working in the field. Many firms still require the insurance and it might be necessary for you to let these jobs go if you cannot negotiate this with them, no matter how much you can help.

7.5   YOU MUST UNDERSTAND THE COMPANIES' POLITICS

Engineering consulting usually requires considerable political savvy. When you are invited to review equipment that is experiencing failures, making the wrong statements can be disastrous. Meetings where engineers and senior management are all invited to review your comments at the same time should be well thought out. I've been in meetings where it's been recommended that a certain analysis be done. Everyone turned and looked at an engineer sitting at the table. The company's engineer had told his management he had performed it. This was certainly embarrassing for him and should have been caught in a smaller meeting with only local management involved. It is preferable to send the finished job and calculations for local review and comments before a face-to-face meeting with senior management. This would have allowed you to have asked the engineer how he had performed his analysis. In this way, it would have been obvious that he hadn't.

This is all quite different than what had occurred when you were working full time for a company. You may have been a Chief or Lead Engineer, where your opinions were usually not questioned and you knew management and the folks who were doing the testing, analysis, and operating the equipment. This is not the case when you are a consultant, which makes things quite a bit more difficult. Collecting data from specialist needs to be handled carefully. Since the competence of the person is unknown to you, the data may be flawed. You also have to be careful with what you say because the designer of the equipment that failed may be seated next to you.

Many times a quick review of the data by the consultant may reveal the failure cause immediately because of past experiences. The desire to comment on this should be handled with the greatest of tact. This company's engineering team may have been working on this problem for some time. Coming up with an obvious solution at the meeting would be most harmful for those involved and they would certainly lose credibility. It would be wise for the consultant to keep this to himself and guide the team in this direction so that the solution is developed by the client's engineering team.

Most often when consulting at a plant, the specialists, operators, and technicians you will be talking with will be assembled by a supervisor or manager. This may not be the optimum team for solving this particular problem. That makes it necessary for the consultant to verify all statements made and the data obtained. For example, once a low-frequency vibration was calculated by the consultant but the company specialists said they had taken data and that frequency wasn't present. Reviewing the vibration instrumentation the company specialist had used revealed that it was unsuitable for measuring such low-frequency responses.

When a meeting is assembled, the consultant may recognize that there is no team leader, the amount of data is tremendous and unorganized, new data is being added during the meeting based on the attendees' speculation, and there is no team organization. This may be the first time everyone has met as a team to discuss the failure. This is why formal problem-solving methods are so useful for solving high consequence failures, as these problems can be addressed by the moderator. The Kepner–Tregoe™ method is an excellent example. Use of it depends on the seriousness of the problem, meaning safety, lost production, or legal issues. The method can be costly and time consuming to apply and requires the use of several experienced specialists from the plant for an extended period. For these reasons, it usually requires high-level approval. In such a meeting, the moderator will then become a tool to direct attention to certain areas such as metallurgy, analytical models, maintenance, process, or management. The engineering consultant would be part of the team and would work the area the team requests and present his information back to the team for consideration at a later date. Management of the team would be left to the moderator who might also be a consultant or senior level site specialist trained in the problem-solving technique used.

7.6 DOCUMENTING THE CONSULTING EFFORT

The report developed by the consultant for his portion of the work, meaning the Final Job Report, should have a section at the beginning called a Management Summary. This is a very concise section, not longer than say one page and presents clearly the problem the consultant was to address, the findings in nontechnical terms, one or two possible solutions and the next steps, if any.

The body of the report would be in terms the team would be interested in and can be quite technical, if required.

The report should not contain information that wasn't requested or presented in the Job Proposal. For example, reviewing other parts of a design, and incorporating your opinions on modifications may not be welcome by others. There may be politics you are unfamiliar with, such as liability or safety issues.

Be careful not to lock yourself into a position by making statements such as "this will solve the problem." It may not because you may have been provided with incomplete information. You don't want to provide statements that would make your work legally binding but only presents suggestions and recommendations. Remember no matter how adept you are at solving problems, engineering solutions are still only approximations.

In the event that you are required to comment on your analysis in a court of law, the opposing attorney may bring in engineers who are willing to contradict your work and your qualifications. There is nothing personal intended and this is just the legal system's method of doing business. You should be prepared for this and only provide information you have verified and the truth as you know it.

Being an expert witness on technical issues pays very well, but I have refused to do such work since I don't care for these type confrontations. You have to be able to remain calm under such interrogations and have an exceptional memory. Both of these talents are not my finest traits.

7.7 USEFUL EQUIPMENT FOR A MECHANICAL ENGINEERING CONSULTANT

When working for large companies I've had the luxury of using the metallurgical laboratories and the talent of the engineers and technicians who worked there.

When I retired and went into consulting, these services were no longer available and I contracted this work to laboratories. To expedite analytical model development, the following equipment was purchased that has allowed me to do a certain level of examination.

A zoom stereo-microscope up to 90 power with photographic and measuring capability has been extremely useful. It can be used to examine materials that have failed and with some basic understanding of how parts fail can help define the mode of failure. This can then be compared with the analytical models results. It would be useful for sizing a surface defect such as a scratch or gouge that might have started a crack. "Beach marks" meaning fatigue or impact growth lines would also be easier to examine and could help verify a cause.

A second valuable piece of equipment is a small 20 power loupe like the jewelers use. I keep it connected to my key chain. When in the field or with a client, there is always the need to examine a part carefully and close up. One can't carry a stereo-microscope around.

A third useful piece of equipment is a metal hardness tester. This comes in handy when doing a wear analysis or need quick material properties. It saves time and keeps from having to utilize an outside laboratory.

More detailed examinations such as material compositions and properties such as tensile strength or fracture toughness require the use of laboratories and their scanning electron microscopes (SEMs), tensile machines, and the talent that knows how to use them. Determining corrosion mechanisms are also best left to these materials experts.

For quick vibration data, a handheld vibration measuring instrument is used. It provides displacement, velocity, acceleration, and frequency data and can easily be taken to the job site. The client company usually has vibration and signature analysis equipment for more detailed data.

Other tools in a machinery and equipment consultant's troubleshooting toolbox are stress analysis, fluid flow, and vibration finite element programs as well as the knowledge and experience to simplify problems and to use these tools.

While certainly not complete, these pieces of equipment have helped resolve the cause of many complex failures.

7.8 VERIFYING AN ANALYSIS

Clients can request a consultant to verify an equipment builder's calculations on new equipment usually after something has failed at an early stage.

This might involve an analysis of a piping system, gear box design, or other types of machinery or fixed equipment. Usually, the data supplied by the equipment builder is minimal and when asked for the calculation procedure used the reply may be that they cannot supply it because it's proprietary information.

This can be a cultural thing and some parts of the world are more likely to have this response than others. It's a difficult problem and one has to be extremely tactful to obtain the information. The relationship the project manager has with the equipment builder is usually the key.

What the consultant can do is provide an independent study and if it finds the equipment to be lacking, the project manager can show the analysis to the builder and ask why the differences. Several things may then happen:

- The builder may supply their information, which can then be verified by the consultant.
- The builder may supply data that doesn't provide the information required.
- The builder may not respond at all.

In the case where the information isn't supplied, one of my colleagues suggested that the project manager move up the management structure at the builder's company. Write directly to the builder's senior management as copies to them by way of the project team may never be read. At some point, the loss of future business will rule and a decision will be made.

Experience has shown that when someone realizes you have done an analysis or detailed research, they are more likely to provide the information you require.

8

PRECAUTIONS ON PURCHASING FIRST OF ITS KIND EQUIPMENT

In industry, engineers at some time during their career will be purchasing new equipment, which is sometimes called "The First of Its Kind."

By First of Its Kind it is meant the first built for your company and can also be called Serial Number One. It could mean a uniquely new design for nontraditional processing equipment or a much larger size, higher horsepower, speed, temperatures, pressures, or product properties. This piece of equipment can usually limit or stop production if it fails, so it's crucial that it's designed correctly.

Specialty equipment builders are frequently utilized for newly developed processes and in these cases only their previous designs can be examined. When your process requires larger machines, a dramatic "scale-up" of their traditional equipment into unchartered territory can occur.

> *Example 1*: The author was once involved with a large scale-up of a heated dryer. Smaller units were cracking the rotating disks [1, p. 315]. An analysis indicated that the scaled-up new unit that was about to be put into operation would also fail. The disk flexural bending stress had not been reduced to a proven acceptable level. The unit was shipped back to the manufacturer from Asia to the United States for major modifications. Several other identical units under construction were modified before shipping.

> *Example 2*: A large scale-up of a vibrating conveyor/dryer [1, p. 317] had been installed and was in operation. Excessive cracking of the welds occurred after start-up. A review indicated that the drive link tolerances were excessive, which

Survival Techniques for the Practicing Engineer, First Edition. Anthony Sofronas.
© 2016 John Wiley & Sons, Inc. Published 2016 by John Wiley & Sons, Inc.

caused this 30-ft-long, 8-ft-high unit to twist and bend thus overstressing the welds. The alignment tolerances were acceptable for smaller flexible conveyors but not for this rigid scaled-up version.

Example 3: An extrusion system for a polymer plant that contained several large complex gearboxes had specifications written and during acceptance shop testing the gearing experienced problems. The manufacturer would not provide their calculations, so consultants were asked to review all of the gear unit design by conducting American Gear Manufacture Association gear verification calculations. They were all found to be acceptable and some only marginally so. With additional attention to details during the initial specifications, the life of some of the gearboxes in tooth pitting could have been greatly extended.

Example 4: Sometimes, it's necessary to see what is occurring before writing any design specification or even to determine if you really want the job. The company I was employed by built large diesel engines. They were used on locomotives and could be up to 4,000 horsepower. We were entering the powering of seagoing vessel market. A potential order for repowering a seagoing tug that took barges up and down the Mississippi river was being considered. I was asked to be on the ship for a day trip and note the operation of the old engine system that was to be replaced with ours. The idea was to see the conditions our new engine might be subjected to. After a few hours, the tug hit a log and one of the blades broke off of the propeller. The ship vibrated terribly for the next 4 h until they could deliver the barges. The coupling was smoking so badly that they put a fire hose on it to keep it cool. The crew said this was fairly common. I provided my report to my company's project manager. The specifications and contract was worded so that if these type events occurred, my company wasn't liable for damage done to the engine system. Firsthand observations are the best kind.

The author has analyzed many of these type failures for various companies. The number of failures could have been reduced by considering the following recommendations.

8.1 INITIAL DESIGN SPECIFICATIONS

Become involved with the manufacturer during development of the design specifications. Design modifications may need to be challenged when the manufacturer makes changes to improve the manufacturing production schedule or reduce costs, which may adversely influence the machine's long-term reliability. It's better to find this out in the early review stage rather than after start-up.

8.2 QUESTION EVERYTHING AND UNDERSTAND THE DESIGN

You cannot write a design specification [2] without understanding the equipment. While the manufacturer can explain how it works on smaller versions, they may not know the problems that can occur on scaled-up versions. In design reviews, something that doesn't appear acceptable to an experienced engineer should be questioned. For example, parts that have had failures in smaller designs should be examined in detail. A simplified analytical model can be beneficial in analyzing the scaled-up version.

One approach is to find a good specification that has already been written for similar equipment. It can then be modified where known failures or reliability issues have existed and have been documented. References to National Codes and Standards are also usually good to follow.

Here's another story to illustrate a point. When I was employed by a research firm, I was given the task of addressing a government request. They often do this and the laboratory had received many lucrative research contracts over the years.

This request for a proposal was for a space vehicle, which was to be the first to land on the planet Mars, take samples of the rocks and soil, and analyze them with specialized instrumentation.

Since I was doing experiments and researching drilling operations, the laboratory director sent the proposal to me. Data was provided that gave the power, duration of operation, weight and size, amount per sample, shock loads, and environmental conditions. The most important specification was that the sampler had to take a sample and deposit it in an onboard container. The substance to be sampled was said to have the consistency of sand or nickel. These were pretty extreme properties, but I determined a twist type drilling device would fulfill the requirements, which it did. We didn't win the competition and never knew who did or what was wrong with our method.

Several years later, the mission to Mars was made and I noted the sampler device used was a scoop much like a small shovel. It could never have sampled a metal like nickel. I figured that the contractor who would develop the device had been chosen and given different specifications than we had even as I was working on the proposal. So here is a case where I didn't have the correct specification to develop the product. No one ever said things would be fair. The decision on which contractor to use had most likely been made before the proposal was ever received.

8.3   DOCUMENT ALL CHANGES AND TRUST NO ONE

The final design may not represent what you initially agreed on and you will want to know why and so documentation is important.

It's your company's production losses, so trust no one. Remember if it isn't written down, you have no legal reason to believe it will be done. If someone says a critical change has been made, physically verify that it has.

8.4   ASSIGN RESPONSIBILITIES

Include in the specifications that an experienced manufacturer's representative will be on-site during installation and start-up and be part of the team.

A knowledgeable engineer from your company should follow the job, with the authority to question and make changes from initial design, assembly, shop testing, installation, and start-up.

When equipment is being manufactured in a country where you have not done business before, it's quite possible that the specifications will be agreed on but not followed. For example, you may require a certain code to be followed but if the country is not familiar with the code, they may use one they are familiar with. Likewise, for producing large metal valves and castings that may be porous. The only way I know of preventing such things from happening is by having your company representative present following and reporting on specific tasks that must be done correctly. This can be expensive, but it's much less costly than finding out about it during start-up.

Even with all of these precautions, it would be naïve to think the start-up will be flawless. What should be expected is that no major design flaws or embarrassments have gone undetected. The question that will be asked by the equipment owner if a major failure occurs after start-up is "Why didn't we find this show stopper in the design phase?" This is the type question engineers and project managers don't want to have to answer.

8.5   WHEN THINGS DON'T WORK AS EXPECTED

Good design specifications are based on historical data that is available at the time. It may be your data, the manufacturer's or someone else's. On new designs, there may be no data such as the permissible conveyor tolerances mentioned in Example 2. This is the type data you will want to add to future purchase specifications, so tolerance problems won't reoccur. When this failed in service, it essentially shuts down the plant until it could be repaired. The equipment owner immediately wanted the legal department brought in to determine if the manufacturer could be forced to perform the repairs under warranty. I'm sure they could, but it was recommended that not be done for the following reasons:

- With the case under the proposed litigation, the manufacturer may do nothing thus further delaying the start-up.
- With any future repair work, the manufacturer may be unwilling to help.
- The manufacturer may be unwilling to bid on future installations.
- With the error corrected and included in specifications for future machines from this manufacturer, this type error is unlikely to reappear.

The plan forward was as follows:

- The owner would provide the welders and fitters to make the on-site repairs at the owner's expense.
- The equipment manufacturer would provide the new parts, design changes, shipping, and a field service representative to be on-site for the installation and alignment of the parts and during start-up at the manufacturer's expense.

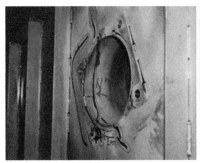

Figure 8.1 Weld cracking of vibrating conveyor.

This plan worked well and at a later date, four new vibrating conveyors were purchased from this manufacturer and worked flawlessly. The manufacturer added tolerance restrictions and vibratory stress limitations to the new specifications, which were verified by strain gauge testing. Note that this was the manufacturer's specification at the owner's recommendation. The owner shouldn't add information to a specification that the manufacturer may not be able to produce, for example, too tight of a tolerance. This agreeable outcome had very favorable pricing and future collaboration on new projects between the owner and the manufacturer.

This type information is important especially on specialized equipment, where there may not be many or any other capable manufacturers. The owner's company may have been in a difficult situation with plant expansions if this had been taken to court.

Several years later, a company in Australia asked for help on a new vibrating conveyor that had been installed and had many cracked welds after a short period of operation. This was a different manufacturer than the one I was familiar with. Some photos were sent that illustrated the failures. Two are shown in Figure 8.1.

Now this was exciting for me because I was asked to go to Australia and help determine the failure cause and work with the manufacturer to address it. I had never been there and was looking forward to the trip and this interesting project. Unfortunately, the company sold the division and the trip was canceled. A colleague heard that the new company scrapped the conveyor. It must have been in pretty poor shape.

REFERENCES

1. Sofronas, A., Analytical Troubleshooting of Process Machinery and Pressure Vessels, John Wiley & Sons, 2006.
2. Bloch, H.P., Geitner, F. K., Maximizing Machinery Uptime, Start with Good Specifications, Elsevier, 2006.

9

USEFUL INFORMATION TO CONSIDER

This section was written to illustrate the importance of recording historical data you have obtained. At some time, your company may have spent considerable resources to record the torque in some system or stain gauge another to determine stress variations. Possibly you have read a technical paper or have done an analysis that provided the data. The following describes why it is important to document them.

Equipment of all types are designed for normal operating conditions, usually with a safety margin to provide for unknowns. The problem is that the potential severity of many loading conditions is unknown.

Every failure is case specific and requires detailed studies; however, when failures do occur it's beneficial to see what magnitude of forces and torques others have experienced on similar failures.

It would be valuable to have strain gauge, load cell, or torque meter data monitoring such failures; however, this is rarely the case. Most data are obtained by observing the failure data and determining by analytical methods what the loads or torques were to cause the failure after verifying the mechanism by metallurgical analysis.

These are cases the writer has experienced for specific failures and the loads and torques could be more or less. However, as a start, in many instances some data is better than no data at all.

Survival Techniques for the Practicing Engineer, First Edition. Anthony Sofronas.
© 2016 John Wiley & Sons, Inc. Published 2016 by John Wiley & Sons, Inc.

9.1 VARIOUS TYPES OF EQUIPMENT AND THEIR FAILURE LOADS

Agitators and Product Mixers:

- When the blades or paddles go in and out of the product and are not fully immersed the mean torque can vary by 3× or more.
- The force on an agitator blade due to product impact, like a chunk of agglomerated rubber in water, can cause the blade load to instantaneously increase by 2× or more over the mean load.

Piping:

- Severe water hammer can increase the reaction loads on piping supports by 4× or more.
- Mixed flow conditions meaning gas with liquid slugs can increase reaction loads, especially if bending moments, by 4× or more.

Extruder and Augers Driven by Motor via Gearbox:

- Torque fluctuations can be ±25% of the mean torque in the gearbox. The use of this piece of information is shown in Section 10.2.
- Blow back in extruders can cause 25% torque fluctuations on the screw.

Centrifuge:

- Depends on the size, speed, and mounting structure. The unbalanced shaking forces due to the product can be 4× the normal operational unbalanced forces.

Torsional Vibration Torques of Systems:

- The magnifier or increase of a fundamental resonance peak over the static forced amplitude can be up to 50×. Most systems were 10× or less.
- When the vibratory torque is equal to or greater than the mean torque on a gear, it can undergo "hammering" of the mesh, meaning torque reversals, which if allowed to continue, can break gear teeth.

Gear Box Tooth Load:

- This depends on what is being driven but could be 2× more than during smooth operating loads. This can adversely influence the pitting life of the gear face.

Reacceleration Loads:

- When power is briefly lost and reapplied to a large generator, the torque to the driven equipment, such as gears, may be 7× or more the mean torque.

Shaft Failures:

- Excessive tension on a "V" belt can actually cause a cantilevered shaft to yield or increase bearing loads by 4× the manufacturers recommended tension.

Synchronous Motor Start Torque:

- The starting torque of a synchronous motor can have a cyclic torque of ±2× or more and will usually smooth out to less than 15% after full speed is reached.

Pumping Pressure Fluctuations:

- The gas pressure fluctuations from positive displacement compressors can be ±20% the mean pressure, while ±5% is common.

Surging Loads in Blower Air Systems:

- That loud "whoosh" surge cycling can raise the blower static pressure by ±25%, while ±5% is common.

Vibrating Conveyors, Augers, and Other Fabricated Machine Fatigue Failures:

- When the cyclic stress on welds is ±5,000 lb/in.^{2} or higher fatigue failures have occurred.
- Impact stresses of over 30,000 lb/in.^{2} can start cracks in fillet welds.

The data presented here can be used to show that a detailed analysis is necessary to help confirm and address a cause, so it doesn't reoccur.

9.2 CRACKING OF WELDS DUE TO CYCLIC STRESSES

9.2.1 Welds with Static Loading

Pressure vessels and piping adhering to recognized welding codes rarely fail in static loading unless damaged. Fabricated structures also are reinforced in critical areas to survive periodic heavy impacts.

Weld cracking can develop in pressure vessels under cyclic conditions [1]. While there has been extensive research and design guides for design of welds under cyclic conditions, the complex metallurgy of a weld and the quality of the weld are difficult to define.

This section is about in-service failures and is based on the author's experience with the fatigue cracking of welded fabrications. The key is that calculations had been done to determine the nominal cyclic stress present when the failure occurred.

In this manner, the limiting cyclic stress could be determined and compared with test data.

9.2.2 Why Welds Fail in Fatigue?

Welds are used to join pieces of metal together. They do this by melting the material and adding it into the melt pool. Metallurgical changes, shrinkage, residual stresses, stress concentrations, internal defects all can combine to cause a much lower endurance limit in fatigue from the base metal.

Fatigue can be thought of as a tensile stress opening and closing a preexisting crack, which may be very small, causing it to grow. In this book, a plus will be used to show a tensile stress cycling from zero to tensile stress, meaning opening a crack.

9.2.3 Weld Fatigue Life

The author is not a metallurgist but has seen and analyzed many in-service fatigue failures. The welds have been all types such as fillet, butt, flush ground, and plug. The loading has been primarily bending.

Figure 9.1 illustrates one of many growing cracks in the toe of fillet welds in a vibrating conveyor that has undergone misalignment and twisting. The failure occurred with only +6,000 lb/in.^{2} on several of the weld within a year's operation. With the problem resolved meaning the cause of the misalignment corrected, the

Figure 9.1 Crack in toe of fillet weld.

stress dropped to $+1,500\,\text{lb/in.}^2$ and the conveyor operated for 10 years without further cracking.

Figure 9.2 shows a failure in a pipe shaft undergoing a bending stress. Notice how the weld starts from one side at a defect and progresses through.

This particular weld in stainless steel had a small initial crack that started to propagate in less than 1,000 cycles at $+30,000\,\text{lb/in.}^2$. This is a good reason to be suspicious when an impact load causes a nominal stress of $+30,000\,\text{lb/in.}^2$ or higher on a weld.

Figure 9.3 is a bending failure in a plug-type weld and starts at an existing design fabrication crack in the "blind" zone that was inaccessible. The weld melt outline is

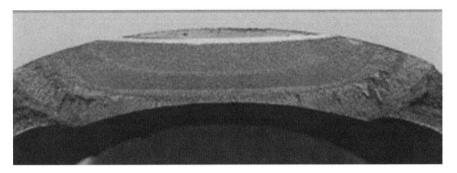

Figure 9.2   Fatigue weld pipe defect.

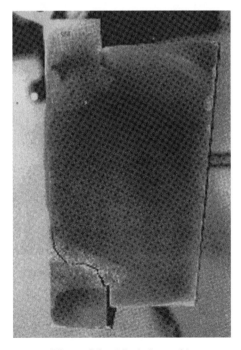

Figure 9.3   Blind plug weld.

visible and was blended flat on the left plate. Notice there is a gap that acts as the start of a crack. Due to this design detail, the nominal bending stress on the plate must be kept low. This was done by using a thicker plate.

All of these failures had a calculated nominal cyclic stress of more than $+6,000 \, lb/in.^2$. Once a fatigue crack starts or exists, it usually continues to grow at these stress levels.

9.2.4   How to Reduce Fatigue Failures of Welds

Designing to keep cyclic weld stresses below those mentioned will help. For example, on vibrating conveyor purchase specifications, the equipment manufacturer agreed that on critical welds the nominal stress should be less than $+3,000 \, lb/in.^2$ as verified with strain gauges. Other techniques are available [2].

9.3   REMEMBER TO CONSIDER ALL FORCES AND MOMENTS

As engineers we see all sorts of machinery fail, so it's important for us to understand as many types of failure mechanisms as we can. This can be enhanced by looking into failures we may not be directly involved with but whose outcome is known. Here's one such case.

A high-performance racing aircraft engine crankshaft had developed a crack in the propeller drive shaft and was being analyzed.

Most of the time stress calculations will be done on this type shaft for stationary machinery. When analyzing machines that are also moving and making rapid turns, another load needs to be considered and that is the gyroscopic force.

A spinning wheel resists movement perpendicular to its axis of rotation. Hold a spinning bicycle wheel by its axles and try to turn it left to right or right to left and it will resist this motion.

The theory of why this occurs is well documented [3, p. 452]. It results in reaction forces on the bearings and shafts of automobiles, aircraft engines, and any spinning accessories when the direction changes from straight line motion. In straight line motion, it's not a factor.

Usually, these reactions are small, but they can't always be neglected as is shown in the following example.

Figure 9.4 is how I visualized the problem when I was first told of it. As with all my visualizations, they are usually quite simple and greatly exaggerated.

An aircraft is making a steep turn. When it's flying straight and level as shown, everything is fine. In the steep climb out, the propeller wants to stay in its original straight and level configuration shown as "A." The steep climb exerts a bending

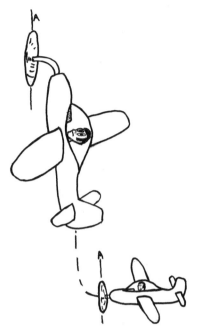

Figure 9.4 Aircraft steep climb.

moment on the shaft due to the gyroscopic effect in addition to the transmitted torque and the shaft tries to bend.

Figure 9.5 puts this into a simple model and equation form. It illustrates a propeller of mass m and diameter D spinning at angular velocity ω_s meaning the angular engine speed and its shaft (d) in a bearing attached to an engine (not completely

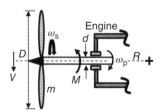

Figure 9.5 Aircraft propeller dynamics.

shown). Also shown is a sudden yaw velocity ω_p when the direction of the aircraft is suddenly changed.

This produces a bending moment (M) on the shaft with a value:

$$M_{\text{ft-lb}} = J\omega_s\omega_p$$

The propeller mass moment of inertia is J, where m is the mass and k is the radius of gyration:

$$J = mk^2$$

Here the propeller will be treated as a thin disk and therefore $k = 0.707 * [D/2]$.
ω_s = angular spin, $\text{rad}/s = 2\pi(\text{rpm})/60$, where rpm is the propeller speed, ω_p = angular precession, $\text{rad}/s = [V_{\text{ft/s}}]/R_{\text{ft}}$ when the aircraft is turning at a velocity V around a circle with radius R, as does a race car in a turn.

The bending stress [4] for a solid shaft with a stress concentration K, due to the gyroscopic bending moment ($M_{\text{ft-lb}}$) only is

$$\sigma_b = 32 * K * (M_{\text{ft-lb}} * 12)/[\pi d^3]\text{lb/in.}^2$$

For our example, $K = 2$, rpm = 3,000, $D = 7$ ft, $d = 3$ in., and $m = 50/32.2$ lb-s^2/ft.
Consider the aircraft is pulling up in a tight circle $R = 500$ ft at 200 miles/h (293 ft/s) after a high-speed pass:

$$M = 1,400\,\text{ft-lb} \quad \text{and} \quad \sigma_b = 12,700\,\text{lb/in.}^2$$

The solid shaft torsional shear stress due to the propeller drive torque alone is

$$S_{ss} = 1 \times 10^6 * K * \text{HP}/(\text{rpm} * \pi * d^3)$$

With 500 horsepower and 3,000 rpm

$$S_{ss} = 3,900\,\text{lb/in.}^2$$

Combining these two stresses into an equivalent stress:

$$S_{\text{combined}} = (S_b{}^2 + 3S_{ss}{}^2)^{0.5} = 14,400\,\text{lb/in.}^2$$

The allowable design stress is 30,000 lb/in.^{2}, so this additional moment and stress is significant.

Under normal maneuvers, the gyroscopic effect is less than 1/4 these values. The reason for the crack was determined not to be due to the gyroscopic force but from other causes. This type calculation would have eliminated this potential failure scenario.

9.4 PHANTOM FAILURES: SOME FAILURES ARE VERY ELUSIVE

While it would be nice to say all troubleshooting efforts have been successes, this wouldn't be true. Those who say they have always found the true cause of failures haven't tried to solve many problems.

There are those failures that haunt us called phantom failures because they are so elusive. Many times we make changes so the problem doesn't reoccur but really haven't found the true causes.

Here are a few examples:

1. A mixer/reactor vibrated excessively; however, when it was opened up and inspected no cause was found.
2. A pipe falls out of the pipe rack and ruptures for no apparent reason.
3. A pipe fails from fatigue at a weld, but there is no vibration in the system.
4. One diaphragm in a steam turbine buckled from excessive force, but there were no apparent operating conditions that could have resulted in such a large force.

The following techniques have been used in the above-listed failures:

1. The mixer/reactor was instrumented for continuous velocity vibration recording to determine at which part of the batch process the vibration occurred. It was during the wash cycle and it was theorized that product had adhered to the vessel wall and was falling off periodically and being chopped by the rotating blades. This caused the vibration. The hot wash oil dissolved the product so that the evidence was gone when the teardown was performed. More frequent cleaning solved the problem.
2. The sudden closure of a valve caused water hammer and the force knocked the pipe out of the rack. An analysis revealed this was possible. Valve closure time was increased.
3. Two-phase flow occurred in the system during operation and the severe slugging caused this remote piping connected to a vessel to vibrate and fail in fatigue. Two-phase flow was not supposed to be possible in this system. The system was redesigned.
4. An incorrect start-up procedure was thought to have been used, although no one admitted to this. Confined water instantaneously vaporized into steam and overpressurized the system. An analytical model revealed this was possible [4, p. 309] so the start-up procedures were modified. This solved the problem as no further failures occurred, but the true cause was never determined.

Thus, for the cases shown here and for other cases, the following approaches are sometimes successful:

• Instrument up the machine or system with vibration, strain gauges, torque, force, displacement, pressure, temperature, oil particle sampling or whatever is required and continuously monitor the results. The hope is to capture the next failure, if one occurs, and acquire the data needed to address it. Unfortunately, the author's experience has been that after a few weeks if the problem hasn't reoccurred, the rented monitoring equipment is removed. It is usually right after this when the next failure occurs, thus no failure data is captured. So keep monitoring as long as practical.

- From the failure analysis, data that has been collected or the analytical model that has been built, address as many of the potential causes as you can. This is sometimes called the "shotgun approach." It's not pretty but is better than not doing anything. If the failure occurs again at least you have eliminated several possible causes. This is one of the major advantages of analytical modeling since many potential causes can be simulated on the computer without disturbing the operation of the unit.

Allowing even minor repeat failures, by simply repairing, because the cause cannot be determined, can often escalate into more serious failures [5]. It is therefore important to fully investigate all critical failures, even the phantom ones.

9.5   THE ART OF HAMMER TAPPING

The reader might well ask, "Why Is He Putting This In?" There are a couple of reasons and the first is that it has been helpful to me because of the practicality and the reader may also find it useful. The second reason is that I thought they were interesting observations put into practice.

It may sound unscientific but a simple "tap" with a small hammer can provide valuable data. After 48 years of engineering along with rebuilding all types of machinery, I am a believer in this subjective nondestructive testing method (NDT). The technique is not in much use today and certainly shouldn't be relied on but as you will see it is a useful tool to know.

Many of us know that when we go to our physician, the doctor hits our knee to check our reflexes. This also verifies the function of certain nerve pathways to our brain. Taps on our body will determine if there is fluid in our lungs and the soundness of certain organs. This was an important part of data taking before modern examination techniques took over such as MRI, X-ray, and EKGs. There were good practical reasons for these taps. They didn't fix anything, but they provided valuable data to help evaluate the patient's condition and they didn't cost much to do.

With machinery, pressure vessels, piping, and structures, this is similar. For example, if a grouping of similar bolts are holding a critical part, "tapping" on the bolt head or stud can tell the experienced investigator if a bolt is loose or broken. The loose or broken bolt will have a different sound when struck because the sound wave return path is different. An experienced troubleshooter will notice this. If nine of the bolts in the group "ping" and one of the bolts "thud," it's probably worth checking further. This method was once used with a 16-oz. ball peen hammer to check the tightness of internal frame bolts on a large gas engine compressor that was having crankcase explosions. Five out of 160 studs were found loose, which caused a hot spot rub on a piston skirt.

This procedure can also identify internal thinning or corrosion on metal tanks and piping since it doesn't "ring" like new metal plates. The soundness of welds was tested in a similar way as were rivets and railroad train wheels. These type checks were actually in some national pressure vessel, ship building, and structural codes pre-1960. Voids, defects, delaminations, and cracks in nonmetals such as concrete foundations or composites can also be subjectively detected. Much like a cracked bell, they just don't ring true.

Today there are electronic devices to do this work. When no equipment is available, these techniques are still useful. Large areas can be tested in a short period of time and then suspicious areas can be examined by other methods. I use to practice by running tests on different broken components to "tune" my ear.

An engineering colleague, who was also a skilled musician, was tapping on a machine housing. He could determine the natural frequency of a part by striking it and noticing the note. He would then look up the frequency in cycles per second for the note. For example, in this case he said it was an "A" (440 cycles/s). More detailed analysis would prove him close to the predominant measured frequency.

"Tapping" can also temporarily fix things. Now we are talking about using a small 6-oz ball peen hammer with a gentle strike not a swing.

Many times I've tapped on a carburetor in a vehicle or small lawn mower engine only to have them come to life with the next start attempt. An observer will usually question your sanity or skill level, but there is an explanation. Carburetors have a float and needle valve to admit fuel. With disuse, the valve becomes gummy from stale fuel and can stick and starve the engine. The hard rap can jolt it free again.

Gentle raps on generators, alternators, or battery terminals have also been successful. Most of the time, the reason was corrosion, loose connections, or stuck brushes.

So the next time you see someone tapping on an object gently, some scientific data taking may be in progress. When you see someone swinging wildly and hard with a huge hammer, they probably are angry and just don't know what they are doing.

9.6   DEVELOPMENT OF SOME SIMPLE ENERGY EQUATIONS

Energy balances are used extensively when developing simplified models and it is worthwhile to see how they were derived if you use them. They are used in many forms such as in heat transfer and also in force and impact calculations such as in Chapter 10.

The development of these equations was not included in previous books [4, 6]; however, since there was interest they are included here.

Two equations are discussed and are the momentum and impulse equations and how they are converted into the average force for use in failure investigations.

The momentum of a mass is the mass times its change in velocity:

$$\text{Momentum} = m * \Delta V$$

The impulse is the product of the average force and a change in time:

$$\text{Impulse} = F_{avg} * \Delta t$$

Newton's second law states the average force on a mass is equal to its mass times the acceleration on the mass, meaning push something with a certain acceleration or try to stop it and some average force will be required.

$$F_{avg} = m * a$$

This can also be written as

$$F_{avg} = m * \Delta V / \Delta t \quad \text{where} \quad m = W/g$$

This then is the average force developed on the mass due to a change in velocity over an increment of time. With these two quantities known, the average force is determined.

Rearranging the equations and this is basically saying that the impulse is equal to the change in momentum or

$$F_{avg} * \Delta t = m * \Delta V$$

Another form of this is due to a deformation process, where something is deformed a certain amount. Now the work done on an object is equal to the kinetic energy applied to the object. Like an automobile hitting a wall, the kinetic energy (KE) is converted into potential energy (PE) of the deformed automobile simply because one form of energy is converted into another form. Heat energy is assumed negligible.

$$KE = PE$$
$$^1/_2 m * V^2 = F_{avg} * d$$
$$F_{avg} = (m/2d) * V^2$$

This is the average force due to the deformation d on mass m with a velocity of V and this force can now be used to analyze loads and stresses.

Many times the spring rate is known and the potential energy (PE) of a spring is required to equate to another form of energy for model development. For a constant force, the PE is the force moving through a distance. This is shown in Figure 9.6 as the area $PE = A_{constant} = F * \delta$.

The potential energy of a spring is $PE = A_{spring} = ^1/_2 F * \delta$.

However, the force varies linearly with δ so $F = k * \delta$, where k is the spring constant in units of pounds per inch.

Therefore, for a spring the potential energy is

$$PE = ^1/_2 k * \delta^2$$

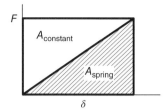

Figure 9.6   Defining spring potential energy.

In Section 10.10, the energy in a volume of compressed air and compressed water are compared and the following equations are used:

For air of volume V in^3 initial pressure $p_1 = 14.7 \, \text{lb/in.}^2$ and final pressure $p_2 =$ gauge $+ \, 14.7 \, \text{lb/in.}^2$.

Potential energy air $[7] = 0.2^* p_2^* V^* (1 - (p_1/p_2)^{0.286}) \, \text{ft-lb}$.

For water of volume V in cubic inches, initial pressure $p = $ gauge lb/in.^2.

Potential energy of water $[8] = p^2 * V / 7.5 \times 10^6 \, \text{ft-lb}$.

This explains the loud bang when a balloon is pricked. A 10-in.-diameter balloon filled with air at 0.3 psig possesses about 9 ft-lb of energy and one filled with water has negligible energy. This air potential energy is converted into kinetic energy and into sound waves, thus the bang and flying balloon pieces.

9.7   MAINTAINING PROFICIENCY IN YOUR ANALYTICAL ABILITIES

In Sections 2.9 and 11.4, methods to continually educate yourself on engineering subjects and methods are shown. In this section, how to keep your analytical modeling skills sharp is discussed and as you will see it's practice, practice, practice.

Throughout my career, I've used analysis to help solve engineering problems. To perform an engineering analysis, you must stay proficient. It's much like training for a sport or on musical instruments. When you don't keep practicing, you will lose your edge when it's required.

This is also true when developing mathematical models. When something fails and you haven't performed an analysis in years and are asked to, you may have a difficult time. Most likely, you will have someone else do this type work, which is not as beneficial or fulfilling to your career.

I sometimes practice by asking myself questions on how something works or why something has gone wrong. In this way, I develop different solutions and also do interesting research in the process.

Here are some examples that have helped keep my brain active. The techniques developed can usually be used later for actual problems in some other form.

Example 1: How hard does a baseball impact a bat?

I always wondered this even before I was an engineer. Assume a baseball is thrown at 90 mph (132 fps) and the bat tip is also moving at 90 mph in the opposite direction.

A standard baseball weighs 5.125 oz (0.32 lb) and from high-speed photography the contact time is about 0.0007 s.

Using Newton's second law,

$$F_{avg} = (W/g) * (V_{ball} - (-V_{bat}))/\Delta t = (0.32/32.2)(132 + 132)/(0.0007)$$
$$= 3,750 \text{lb force}$$

This type analysis was used on analyzing the impact force of product on a mixer blade.

Example 2: Why doesn't the large baseball impact force with the bat break the batters' arm?

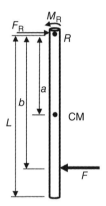

Figure 9.7   Bat impact.

There is a position on the bat called the center of percussion, b. When a ball strikes here batters will feel very little in their hands and wrists.

For simplicity, consider a round bat with hands at R. Bats are heavier at the end where the ball strikes. Since I was not interested in an exact answer, the round bat will do.

So the question I was interested in answering is where should the ball be hit so no force is felt on the batters' hand.

The reason a round bat was selected is because the polar moment of inertia (I) is easy to calculate.

We want the value at R, so we will use the parallel axis theorem to get it there from the center of gravity (CG), where m is the mass of the bat (Figure 9.7).

$$I_R = m * L^2/12 + m * L^2/4 = m * L^2/3$$

The moment at point R for a stick mass (m) trying to rotate about R:

$\Sigma M_R = (m * L^2/3) * \alpha$, where α is the angular acceleration of the bat at R.

Now α can be converted into a linear acceleration acting at CM:

$$\alpha = a_{CG}/a$$

Also,

$\Sigma M_R = F * b$

Using Newton's second law and summing the moments,

$$F_R + F * b = (m * L^2/3a) * a_{CG}$$

Because no reaction at the hands is desired $F_R = 0$, and the translation force is $F = m * a_{CG}$.

Solving for the center of percussion,

$$b = (2/3) * L$$

This indicates the ball should hit about 1/3 from the tip for this straight stick. This would be about 11 in. from the stick tip for a stick length of $L = 34$ in. This is considerably different from an actual bat, where the "sweet spot" is about 7 in. from the heavy end. Feeling the bat's vibration is different as this analysis only looks at the force elimination due to instantaneous contact.

I can personally attest to the fact that hitting the ball at the "sweet spot" will not make you a good hitter. There's a lot more to it than that.

Example 3: Why did my car windshield break from a small stone dropped from a sand truck?

A 1/2-in.-diameter d (0.04 ft) stone weighing W (0.007 lb) falls off a truck traveling at 60 mph (88 fps) and you are following at the same speed. When it hits the ground it sort of stops and bounces up to your windshield level and then hits with about the same velocity.

Simplify by equating kinetic to potential energy since one energy is converted into another form, where $F * d$ is the work done by force F in distance d:

$$1/2 W/g \ V^2 = F * d$$

Solving for F of impact is

$$F = (1/2) * (0.007/32.2) * (88^2)/0.04 = 20 \, lb$$

It doesn't sound like much, but it all depends on the contact zone ($A_{contact}$), which could be due to a sharp corner on the stone.

Consider the contact stress of pitting glass enough to start a crack. My windshield had a pit diameter and depth of about 1/32 in. and a crack had started from it.

$$\sigma_{contact} = F/A_{contact} = 20/(0.785 * (1/32)^2)$$

Stress-wise it's about 26,000 lb/in.^{2}, which is in the glass fracture range. Hitting a windshield with a pointed hammer and light swing proves this point.

Now I try to pass trucks carrying loose gravel as soon as I can.

Example 4: What speed will a locomotive in a curve roll off its tracks?

I questioned this after a series of recent fatal railway accidents occurred on curves.

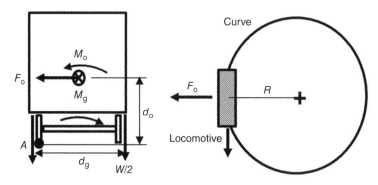

Figure 9.8   Train tipping.

Consider a locomotive travelling around a curve of radius R. At what speed will it roll onto its side? (Figure 9.8).

A locomotive with a velocity V around a flat curve, meaning not banked, of radius R produces a centrifugal acceleration force in the F_o direction, like swinging a ball on a string:

$$F_o = W/g * a = (W/g) * (V^2/R)$$

$$\Sigma M_A = 0$$

$$F_o * d_o - (W/2) * d_g = 0$$

Solving for the velocity to tip the locomotive,

$$V = (g * R * d_g/2 * d_o)^{1/2}$$

Assume $g = 32.2$ ft/s^2, $R = 1,000$ ft, $d_g = 4.7$ ft, $d_o = 4$ ft, then $V = 138$ ft/s (94 mph) and over that it will probably tip.

When a train is going 106 mph in that curve, you know it will probably go on its side.

High-speed trains going 220 mph are designed with $R = 12,000$ ft. Tipping for them with the assumptions used in this example equation is $V = 325$ mph, which provides a nice margin from tipping.

These are more of many brain exercises I have considered and ran calculations on over the years.

- How many hours does it take to heat up a steel part in a furnace?
- What is the speed of wind to knock someone over?
- What is the flooding water flow to move a house?
- What force does an air bag hit someone with?
- What force does a bird hit a small aircraft with?

- What size air-conditioning unit do I need for my garage?
- Why did the Shuttle Challenger "O" rings fail?

9.8 SAFETY CONCERNS TO BE AWARE OF

Most machines and pressure vessels are designed with safety in mind, but from their very nature they can release large amounts of energy if mistreated or operated inappropriately. Here are some cautions I've learned during my career, some from analysis and others from news reports. There are many more available [9].

9.8.1 Pneumatic Testing of Piping and Pressure Vessels

Pneumatic testing is normally performed for leak detection at low pressures and small volumes when water weight or disposal is a concern. Hydrotesting is performed at higher pressures for leak testing, advantageous stress distribution, and structural integrity.

We still see catastrophic failures when pneumatic tests of large volumes are performed [10, 11]. Just like a balloon filled with air and one filled with water, when pricked with a pin one goes "bang" and the other just leaks. There's a lot more energy in compressed air than water. A vessel pressured to 50 psig with air and a volume of 200 ft^3 is about 14,000 times more or about the same energy as 1 lb of TNT. Saying this another way, it's the same energy as a truck going 75 mph. So when someone at a project meeting says, "It's going to contain gas at that pressure anyway so why not pneumatically test it?" remember to speak up. Flying fragments can go thousands of feet. Codes and Standards may only be considering safe distances due to the pressure wave not the flying fragments. There are other ways to pressure test as discussed in Section 10.10. The equations used for the contained energy are shown in Section 9.6.

9.8.2 Reciprocating Machinery

I remember my first visit to a site with hypercompressors, where discharge pressures can be higher than 20,000 psi. I was shown the bent beams in the roof, where the compression end plungers had come apart and pieces had impacted the beam due to improper operating procedures.

Lower pressure gas engine compressors have their concerns too. I have seen rods come through the sides of the engine, and cylinder heads blown off due to a lack of proper preventative maintenance or operation.

9.8.3 Rotating Machinery

One purpose of the housing on machines is to keep the fluids and parts contained if the rotating parts such as blades come loose. Aircraft engines swallowing birds or hail are examples. When blades come loose, they can go through the engine cowling with ease.

Turbocharger rotors large and small and clutch discs can also break through their housing. I've witnessed a turbocharger test in a test cell, where the rotor broke out of the housing and bounced off the test cell wall.

9.8.4 Pressure Vessels and Brittle Fractures

Brittle fracture described in its simplest form is the rapid extension of a crack in a low toughness steel at stresses well below yield. The crack may grow at a rate of 7,000 fps [12] without warning. At this speed, there's no value in trying to monitor a crack. My advice has always been that if an old low toughness steel vessel has a growing crack, don't try to monitor it, shut it down immediately. Replacing such brittle old vessels is usually a wise choice.

Luckily, my only experience witnessing a brittle fracture is shown in Figure 9.9. I was at a shop inspecting a vessel when I heard a loud "bang" outside. We all went out to see what had occurred and there was the end of the heat exchanger shown from where it should have been to where it was located. The technician said he was hydrotesting it after a repair and then he just heard the bang and saw the part fall to the ground. Too cold of a temperature for the steel, a weld defect, and the 50 psig hydrotest pressure caused the problem in this low toughness steel. Fortunately, the test was a hydrotest and not a pneumatic test. The hydrotest had less than 4 ft-lb energy and a similar pneumatic test would have had 25,000 ft-lb.

9.8.5 Welded Pressure Vessels Under Cyclic Loading

Depressuring and repressuring many times per day is cyclic loading. This can induce a cyclic stress on poor welds or overlapping fillet welded plates making them

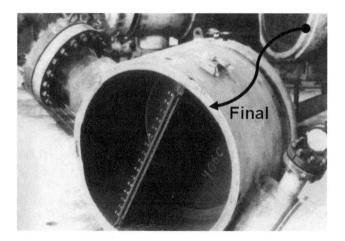

Figure 9.9 Brittle fracture heat exchanger head.

susceptible to a fatigue failure [1]. Good designs, inspections, and sound welds can minimize this risk. An analysis by a reputable engineering firm can determine if this is a problem with cyclic equipment.

9.8.6   Valves Handling Hazardous Material

Valve selection should suit the type fluid being transported. A rubber-lined valve that remains closed for years probably won't operate correctly when needed. It's like keeping your car parked on a rubber garden hose for years. When you move it and turn on the water pressure it will crack and leak. Valve maintenance is imperative especially when people are working nearby. Initial selection may not have been done correctly, so it's a good idea to review the internals when in a critical service.

9.8.7   Plant Steam Leaks

Steam from boilers can reach 2,400–4,000 psi. Any leaks from superheated steam would be invisible due to a lack of condensate. The temperature could be in the 1,000 °F range. I had been cautioned on this as a young engineer by experienced veterans when working at sites with these pressures. The comment was to stay away from all potential leak sources such as split lines, flanges, corrosion points, or valve stems. That was nice to know even for lower pressure steam, but the problem was you don't usually know where they are. So stay away, be aware, listen to the old timers' advice, and always have an escape route if something fails. Even a ruptured steam hose can cause some terrible burns.

9.8.8   Plant Fires

Any time there is a fire in a plant, there is a chance for a massive explosion. For example, consider the poor repair weld patch that failed in cyclic service after several years on a 5,700 ft^3 hydrocarbon vessel. A fire and explosion occurred that sent a 120 lb steel piece flying 900 ft into a community. It moved the 25 ton vessel 150 ft [13]. A Code repair weld would have prevented this failure.

9.8.9   Hydraulically Fitted Couplings

All couplings have the potential for coming apart when rotating and coupling guards should be robust and always in place. Static failures also can occur. On hubs removed hydraulically, the hub can "pop off" at over 25 mph and travel a distance of many feet [4]. Be sure to use a safety nut and most importantly treat the hub as you would a loaded gun and stay out of the line of fire.

9.8.10   Some Recommendations

Always be aware of your surroundings when near equipment. Know the normal operating sounds and odors and understanding that bad things can happen.

Loud knocks, bangs, rapping sounds or hissing, squealing, buzzing, or an unusual sensation in the soles of your shoes may signal something has changed. Something has come loose, the process is in distress, or something might be vibrating or leaking. When you touch a pipe or vessel and it feels like a snake squirming in your hand, the surface is probably in some mode of plate vibration.

Smells are important too. However, when you don't smell something it might be too late. Hydrogen sulfide is an example. In deadly concentrations, it can deaden the sense of smell and isn't a reliable indicator of that "rotten egg" smell.

It's also a good idea to remember that when there's a fire to run in the opposite direction the firemen are going. They are going to the fire.

9.9   SHOULD I PURSUE A PATENT?

I have a notebook full of things I'd like to build and think are needed. I sketched up a device that would bale fallen leaves from trees so they would be easier to dispose of. This came about when I was raking up piles of them. Another thought came when I burned my hand on an exhaust manifold. I used thermocouple devices to generate electricity from the waste heat. Also I note that I had a design for a thermoelectric air-conditioning unit with no moving parts when none were on the market. The motorized aircraft tow, aircraft vibration monitor, and the brain saving bubble wrap cap have all been mentioned in this notebook.

As engineers that's what many of us do. We're always trying to come up with better ways for doing something or fulfilling a need. During my brief period as manager of advanced engineering, my group had to come up with ways to utilize the company's many products. It's an extremely difficult thing to do and fortunately there are various brainstorming techniques available to help a team do this.

I'm often asked about an idea a friend has come up with. It's usually something like, "I have this idea, what do you think of it?" After you have reviewed it and say you think it's a good idea, the next question they ask is how do they patent it.

I usually explain by using my experiences. I have no patents as they have all been filed by the companies I worked for and are their property, as agreed on in my hiring contract.

I have looked into a patent for the motorized aircraft tow and found out it's very expensive to apply and receive a patent and it takes a long time. You see someone has to review the records, usually a patent attorney, to see if it's patentable and then it has to be written up with drawings and explanations in the normal bureaucratic way. This takes time, tens of thousands of dollars, and patience. Filing for a patent is less than $1,000 and it's obtaining one after filing for it that's costly.

Some call a patent a license to steal. What do you do if someone designs and sells what you have a patent on with a slight modification made? You can take them to court to see if there is an infringement on your patent but even if you win it will cost you money. What if someone in another country builds it and sells it?

I have seen one of my seminars presented in a country I've never been to. It said I had presented it. Someone took my seminar handouts and made a presentation out of it. What can I do about that? Just smile and be pleased someone is benefiting from my work I guess.

As for my motorized aircraft tow that I no longer build, I now see other similar ones on the market and I'm glad I didn't try to patent mine. All that money would have been lost.

There are companies that will pursue a patent for your product and also build, market, and sell it for you. They will definitely make money from you; however, you may lose some. You need to know the market and the finances to make a profit.

Sometimes it's better to build it without a patent and see what happens. That's what many companies do. They have their legal staff deal with problems and infringements when they come up. They have a lot of money to do this.

REFERENCES

1. Sofronas, A., Fitzgerald, B., Harding, E., The Effects of Manufacturing Tolerances on Pressure Vessels in High Cycle Service, A.S.M.E., PVP Vol. 347, 1997.
2. Maddox, S.J., Fatigue Strength of Welded Structures, Abington Publishing, 1991.
3. Holowenko, A.R., Dynamics of Machinery, Wiley & Sons, Inc. 1955.
4. Sofronas, A., Analytical Troubleshooting of Process Machinery and Pressure Vessels, John Wiley & Sons, 2006.
5. Bloch, K., Extreme Failure Analysis: Never Again a Repeat Failure, Hydrocarbon Processing Magazine, April 2009.
6. Sofronas, A., Case Histories in Vibration and Metal Fatigue for the Practicing Engineer, John Wiley & Sons, 2012.
7. Coleman, M., Energy Release From Failure of Pneumatic Vessels, ESMC-TR-88-03, Unclassified, General Physics Corp, Florida, 1988.
8. Streeter, V.L., Fluid Mechanics, 3rd edition, McGraw-Hill, 1962.
9. Kletz, T.A., What Went Wrong: Case Histories of Process Plant Disasters, 4th edition, Elsevier,1999.
10. Pneumatic Test Explosion in Shanghai LNG Terminal, Chemical & Process Technology, March 2009.
11. Pneumatic Test Failure in Mississippi Pipeline Project, July 2009.
12. Barsom, J.M., Rolfe, S.T., Fracture and Fatigue Control in Structures, 2nd edition, Prentice-Hall, 1977.
13. DOE/EH-0699, Explosion and Fire Texas Chemical Plant, June 2006.

10

CASE HISTORIES USING ANALYTICAL MODELS

This section provides a sampling of technical problems I have worked during my career. This portion is directed toward what a mechanical engineer might actually do using analytical modeling techniques. It shows how the unknown can become known.

In most cases, such models don't have to be extremely accurate, but they do have to be believable. They are used to explain failures or weak points in designs and since they are relatively simple, they are verified with test data from the literature or from actual operation. Due to their simplicity, they usually contain many assumptions and with those assumptions comes the need to verify the results. How to do this is explained in this chapter.

Mechanical engineering is a very broad discipline and there are so many facets that no book or person could cover them all. One person's career can only be devoted to a few areas and even those few will only be partially understood.

For example, I know of someone who worked for an antifriction bearing company and spent his whole career on rolling element bearings. Not the machines they were used in just the design of the bearings. This engineer did grease testing, design of bearing load testing machines, metallurgical examination, improve steels for the races and rolling elements, stress analysis on bearings using advanced stress analysis tools, and investigations on the cause of failed bearings. Now there are many different types of bearings in industry, but he was an authority on only this type bearing. When he retired, he told me that he had only begun to learn about these types of bearings. A humble and truthful man!

We have discussed niches. Bearings were the retired fellow's niche and passion and he made a fine career of it and consulted and taught on the subject after he retired. He

Survival Techniques for the Practicing Engineer, First Edition. Anthony Sofronas.
© 2016 John Wiley & Sons, Inc. Published 2016 by John Wiley & Sons, Inc.

was able to put his children through college, live in a fine home, took trips around the world with his wife, and retired with a comfortable savings account. He is a happy man because he had a niche and a passion for engineering and enjoys his work and his life.

After 48 years in mechanical engineering, I totally understand him. I have spent very little of my time on bearings but worked on many types of machines that contain bearings. This brings me back to the reality of how much I really know about my field, which is very little.

I have a passion for engineering as you can probably surmise as you read this book. I have had a niche too that I developed along the way. It's an area all engineers do in some form, but I took it to an extreme in my career and that is analytical modeling of machines.

In its simplest sense, this involves looking at a piece of machinery like an automotive engine or the gearbox in a ship and simplifying it into mathematical equations. These are equations that determine the forces, displacement, and stress on the structure due to dynamics, vibration, heat, pressure, and other factors. When the modeling is done, you can now analyze what changes in the loads such as changing the horsepower, speed, temperatures, or pressures have on the system. You are basically seeing what is happening inside the machine while sitting at your computer. Think of what this means! You can determine if something is going to fail, when it's going to fail and why it's going to fail and all without speculation, guessing, or testing. The machine may not even have been built yet.

This has been a powerful tool during my career. The beautiful part of this approach is its simplicity and how it can be used to explain difficult failures to management, something all engineers are expected to do. Albert Einstein once said, "If you can't explain it in simple terms, you don't understand it well enough." I have found this to be quite true when building analytical models.

The following sections give you a flavor of what a mechanical engineer working on machinery may do. These are the type jobs I was responsible for in the problem-solving effort. There were many other talented personnel involved in the solution and implementation of the final remedy. Machinists, operators, inspectors, metallurgists, technicians, engineers, managers, supervisors, and many others all played a part and each had their special talents. It's always a team effort.

Usually, the company you work for doesn't really care how you solved the problem only that you did. Thus, those rather detailed analysis you developed and are quite proud of can go unnoticed. You can get your gratification in that it allowed you to understand the failure mechanism better than others and to make major contributions in solving the problem. With an accurate analytical model built, you are not intimidated by anyone not even the manufacturer of the equipment. In some cases, you may understand the failure cause better than the original equipment manufacturer (OEM). That will be noticed by the management of your company.

It's wise to document the details of your analysis because at some point in your career you may see a similar failure or you may want to write a technical paper or at some point even a book.

Building a complex model may result in a more accurate and eloquent solution, but a prime requisite for engineers is to solve the problem, implement a solution, and be able to explain what you did to those whose expertise may not be interpreting complex analysis. A less complex model can do this. For example, a three-dimensional finite element model will usually be more accurate, but a two-dimensional model will be quicker to build, change, and understand. Such a model will require the engineer to be innovative and have a better understanding of the equipment to simplify the model.

Some analyses were done on high-profile failures, meaning those involving safety and liability but most were on high production loss failures. Many times a simple model was built, which indicated that a more detailed and costly one was justified.

Building analytical models requires the investigator to research other failures and determine the type of failure that has occurred. Was it a sudden failure, a cyclic fatigue failure, bending, torsion, or some other type? The analytical model must agree with the type of failure that has occurred.

As is often said, "Good calculations are always better than speculation," since they can usually provide confirming data.

Closed form solutions can be more versatile than building a new finite element model for every geometry change. Assume you are analyzing the stresses for a press on fit due to a temperature change for troubleshooting purposes. With one model on a spreadsheet, you can just change the geometry or other variables to determine the stresses and obtain a solution. Many sensitivity checks can be run at minimum time and expense. No need to develop and run a finite element model every time.

Sections 10.1–10.4 put this all together and shows you my thinking during the development of the model on a failed piece of equipment. It shows the value of such a model.

10.1 BUILDING AN ANALYTICAL MODEL OF A MATERIAL PROCESSOR

This case represents an important method that I have used for analyzing machines. In this case history, a complete although summarized analysis of the failure of a machine is presented and is a good illustration of analytical modeling, the thinking process that goes into it, and what such an analysis can do. It shows how you must be knowledgeable about the machine and also how to simplify something that is fairly complex. The important point is that the failure area is known by seeing a large crack.

10.1.1 Why Build a Model?

The analytical modeling of machinery has proven to be valuable in the processing industry. When a failure occurs on a piece of equipment, the failure type and location is directing the investigator to the cause. The investigator must then make sense of all the data obtained and fit it together to find the true cause or causes and address them so that the failure doesn't reoccur.

Data collection, discussion with knowledgeable parties, historically similar failures, and a metallurgical analysis of the material and failure zone are all valuable inputs.

When the loads on the equipment are known, they can be used to examine the stresses and deflections of the system both mechanically and thermally. Bearing loads, weld cracking and life, mount loosening, vibration, and distortions can be evaluated.

The manufacturer of a machine doesn't provide this type information and usually call it proprietary information. Most of the time it isn't and it is just an engineering analysis and that's where analytical models can help.

In this case, a product recycle chopping machine used to reduce 75 lb bales of product into small pieces of material for molding presses, sheets, or extrusions is analytically modeled. The desire is to determine the forces and moments on the structure so that they can be used in analyzing the failure zone and why the failure cracking occurred. This was an interesting failure since several failures on similar equipment had been observed. All were not analyzed, but the life before failures occurred was long and was in the 20–30 year range. This would usually be considered a respectable life for a welded structure under cyclic loading. What was desired was to know what would happen to the life of a newly installed chopper if the production rate were increased or the bale properties were changed. Would the long life the original machine had experienced be reduced? What do you do when the manufacturer says it will be fine? You can take the manufacturer at their word or ask for the calculations or test data that show its life is acceptable. When they will not supply the calculations, you can have someone run them for you. When a new machine has been put into production with conditions more severe than anticipated, will the life be shortened significantly? The equipment owners will certainly be upset if a similar failure occurs in a few of years on their new machine. By knowing this early, this might be avoided by changing certain operating parameters.

10.1.2   The Modeling Process

When modeling a machine, the engineer needs to include all information that can affect the loads and moments on the machine. For the case of an auger type chopper that is chopping a bale and compacting the material through a die, this includes the horsepower, auger speed, machine and die geometry, mountings and shear properties of the material along with the areas in shear and the production rate meaning bales processed each day.

10.1.3   Modeling the Chopper Loads

First the machine has to be simplified into a model that includes all of the important variables. When a model is built but nothing can be changed to improve the situation, it is not a worthwhile model. For example, changing the horsepower, speed, screw type, feed rate of a bale being processed, type material being cut, or die hole size by intuition will change the loadings and reactions on the machine and thus the machine life. How much and in which way will need to be determined.

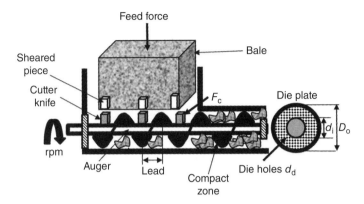

Figure 10.1   Cutter model.

Figure 10.1 represents a sketch of the simplified chopper. A bale of product is shown being forced on the auger–cutter knives. On some machines, this is done with a hydraulically assisted ram. In that case, the force due to the ram would have to be added to the cutting forces. This unit didn't have such an assist and the bale weight alone provided the force. The cutter knives shear the material into chunks that are transported to the die via the auger screw. The harder the bale is forced, the deeper the knife cuts. In the auger chamber section, the frictionally heated product pieces are forced through the die holes by knife augers. I pictured the operation like the old hand-operated hamburger processing equipment mashing up the meat and pushing it through a die plate. Since the product composition can vary, it is obvious that the material properties are important and harder materials will develop larger forces. Thus, the need to have the shear failure stress of the material as a variable. The screw design can also be changed as can the cutter knives and number so they are included in the model too.

10.1.4   Development of Equations

The horsepower available and dissipated by the drive motor to power the unit is

$$HP_{total} = HP_{cutter} + HP_{compaction} + HP_{friction\ heat}$$

By calculating each of these, a rough check can be made on the adequacy of some parts of the model since the rated horsepower HP_{total} is known. The frictional heat is caused by the continual kneading of the product until it is extruded through the die. Since it does soften and heats up the product by the kneading action at the die, the product temperature must be considered when running the shear tests.

10.1.5   The Cutter Horsepower HP_{cutter}

The force required for a knife cutter to shear a piece of product from the bale is

$$F_c = S_s * A_c \, lb$$

TABLE 10.1   Shear Test Results at Temperature

Bale Temperature (°F)	S_s (lb/in.^{2})	Cut Temperature (°F)	S_{sp} (lb/in.^{2})
80	150	160	25

These cutter forces result in reaction loads and moments on the machine.

In the equation, S_s is the shear stress needed to shear a chunk of the product in pounds per square inch and is determined from test data as shown in Figure 10.2 and Table 10.1. A_c is the area that the chunk is sheared out in square inches and is determined by estimating or looking at a piece and measuring the sheared area. One way of estimating this area for a knife of length (l) is that each knife cuts out a shear area of approximately two triangles meaning $A_c \approx l^2$ as shown in Figure 10.3. Since the way the bale will fall on the cutter is random, the exact value isn't possible and things will be kept simple.

The cutter horsepower is

$$T_{\text{in.-lb}} = F_c * N_c * D_o/2$$

$$\text{HP}_{\text{cutter}} = T_{\text{in.-lb}} * \text{rpm}/63,000$$

N_c is the number of cutter blades, rpm is the auger speed, and D_o is the auger diameter.

Figure 10.2   Shear failure testing machine.

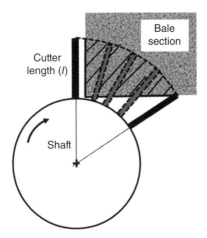

Figure 10.3   Approximate cutter area in shear.

10.1.6   The Compaction Horsepower HP$_{compaction}$

The product is transported via the knife auger to the die face. The flow is due to the processing time of a bale through the auger lead (L in inch) as shown in Figure 10.1.

The shear of the product through the die holes is what causes the axial force on the die and thus force on the structure. It causes the load on the bearings and structural members.

The force required to shear out a plug of product uses the same methodology as the cutter force. In the compaction zone, the chopped product is mechanically pushed through the die holes by a rotating knife in front of the die. The shear stress of the plug S_{sp} is much softer than bale material S_s since it has been chopped and heated by the continuous chopping and kneading action and it will be determined by a shear test at temperature as shown in Figure 10.2.

Figure 10.4 shows a simplification of the force to push a plug of the product through the hole. Here $F_{compaction}$ is the force on the end of the plug, S_{sp} is shear stress on the surface area.

With a die plate with N_h holes of d_d diameter and thickness (t), the axial reaction force is

$$F_{axial} = N_h * \pi * d_d * t * S_{sp}$$

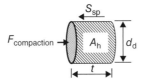

Figure 10.4   Shearing of plug through die hole.

The velocity can be determined by the mass continuity equation along the auger:

$$V = W/(12\rho * A * t)\,\text{ft/min}$$

here W is the bale weight in pounds; ρ, the density in pounds per cubic inch; $A = (\pi/4) * D_o^2 \,\text{in.}^2$; and t, the time in minutes to process a bale.

The compaction horsepower $HP_{\text{compaction}}$ is approximated as

$$HP_{\text{compaction}} = F * V/550 = (F_{\text{axial}} * V\,\text{ft/min})/550$$

10.1.7   The Frictional Horsepower $HP_{\text{frictional}}$

Once the product is chopped, it is continually kneaded as it moves to the die. Some idea on the temperature rise of the product and the horsepower required to cause this rise is available from the specific heat equation [1, p. 166]:

$$q_{\text{Btu/min}} = m_{\text{lb/min}} * C * \Delta T$$

Also recalling that $1\,HP = 42.2\,\text{Btu/min}$, C is the specific heat of the product in British thermal unit per pound-Fahrenheit and m is the weight flow in pounds per minute of the product through the unit. The heat of the product can now be determined.

For example, if a 75 lb can be processed in 5 min, $m = 15\,\text{lb/min}$ and if $C = 0.5$ and the product temperature difference out versus the product in is 80 F, then

$$HP_{\text{frictional}} \approx 14 \ \text{horsepower}$$

This horsepower does not contribute to cyclic forces causing the failure but does show that not all of the horsepower goes into cutting and pushing the product through the die.

10.1.8   An Example of What Can Be Done with the Model

Assume the following information is available on a machine of the type shown.

Units are inch-pound-minute

$HP_{\text{total}} = 100$ available, rpm $= 49$, $S_s = 150$, $S_{\text{sp}} = 25$, $A_c = 25$, $D_o = 8$, $d_d = 1$, $t_{\text{die}} = 1$, $N_c = 6$, $N_h = 80$.

The axial force pushing material through the die is $F_{\text{axial}} = 1{,}600\,\text{lb}$, and can be used to evaluate the bearings and the force on the die face. The reaction cutting force on each knife is $F_c = 3{,}500\,\text{lb}$, which can be used to examine vibratory forces, bearings, and can also be imposed on the structure so structural stresses and distortions can be calculated. As a rough comparison, the horsepower required by the cutters is 112, the compactor 3, and frictional is 14.

10.1.9   Testing Required

Because of the sensitivity of the model, values of S_s and S_{sp} should be determined by testing. A physical examination and a load/shear failure test on actual processed pieces of the material will be needed. A simple press set up to shear a sample of the material with a known load would provide this information. The shear failure values could then be used in the model. Figure 10.2 illustrates such a setup.

To test in shear, a piece of the material is placed on the cutter and a known force applied until a shear failure occurs. The shear area is measured and the failure force divided by the sheared area results in the static failure shear stress S_s or S_{sp} in pounds per square inch. The temperature the product exits the die is used in the shear test for S_{sp} and is due to the kneading action on the product after it is cut.

Table 10.1 illustrates test data for the product.

Section 10.2 shows how these loads can be used to estimate the loads on the structure.

10.2   DETERMINING THE LOADS ON THE PROCESSOR STRUCTURE

In Section 10.1, the forces due to cutting the bale have been calculated. Now the forces on the structure have to be determined, so the stresses in the failed area can be determined from the loads.

Figure 10.5 shows the cutter loads along the shaft. These loads also act on the structure.

Figure 10.6 illustrates how the resulting moment M_1 acts on the feed end plate and on the connecting housing. A failure of the weld by cracking occurred at the area of the moment M_1 shown. The location of the mounting feet was very important in

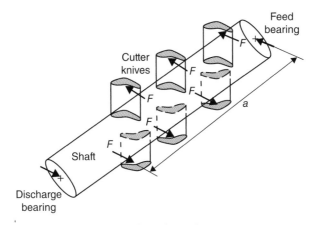

Figure 10.5   Cutter forces.

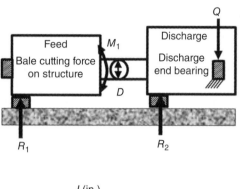

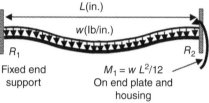

Figure 10.6   Cutter forces on structure.

analyzing this failure. The cutting forces resulted in the moment M_1 on the feed end plates distorting the plate.

In Figure 10.6, Q is the bearing reaction in the discharge section but doesn't significantly affect the moment. Solving for the moment at the failure zone [2, p. 20] results in

$$M_1 = wL^2/12$$

with $N_c = 6$, $F_c = 3,500$, $L = 28$, $w = 6 * 3,500/28 = 750 \, \mathrm{lb/in.}$

$$M_1 = 49,000 \, \text{in.-lb}$$

This moment can be applied to the weld and the end plate and the stresses calculated either by the finite element method (FEM) or plate theory.

$$\sigma_{\text{bending mean}} = 10,000 \, \mathrm{lb/in.}^2$$

Only the cyclic moment affects the fatigue life as will be seen in Section 10.3 and the above stress would be the mean stress. How much of this is cyclic? This is a difficult question to answer without test data, but Section 9.1 suggests an answer.

In extruder and auger type equipment, the output torque can have a fluctuation of $\pm 25\%$ on the mean torque as discussed in Section 9.1. While an auger type cutter is not exactly an extruder or feed auger, it is an assumption that will be made. So to have this stress be cyclic only 25% of it will be considered since it was a fatigue failure.

$$\sigma_{\text{bending cyclic}} = 0.25 * 10,000 = 2,500 \, \mathrm{lb/in.}^2$$

The cyclic portion needs to be considered when fatigue is an issue. When this cyclic stress occurs at welds and failures have occurred, a fatigue analysis or crack growth calculations are necessary. This will be done in Section 10.3.

The axial force pushing on the die plate and pulling on the housing is

$$F_{axial} = N_h * \pi * d_d * t_{die} * S_{sp}$$

with $N_h = 80$, $d_d = 1$, $t_{die} = 1$, $S_{sp} = 25$.

The axial force is

$$F_{axial} = N_h * (\pi * d_d * t_{die} * S_{sp}) = 1,600 \text{ lb}$$

This load is taken up by the support, so it is not transmitted to the cracked end plate.

10.3   DETERMINING THE LIFE OF THE PROCESSOR

The life of a component in years under various operating condition is an important product of an analysis. This is very important in welded structures as the cyclic stresses to propagate cracks can be much lower than the yield strengths of materials.

While we tend to treat a weld as a solid piece of metal, most have surface cracks, slag inclusions, voids, or incomplete welds. How fast this defect will grow into a failure depends on the magnitude of the cyclic stress trying to open a crack and returning it to zero and how many cycles occur.

Here is a procedure for approximating the crack growth [1]:

1. For stainless steel plate with a thumbnail crack $2c/a = 6$
2. Service factor (time run) 365 days year $= 1$, half that time would be 0.5
3. Plate or weld thickness t
4. rpm or cpm vibration
5. a_0 Starting crack depth defect size use 0.0625 in.
6. a_f Final crack depth use t thickness weld or plate
7. $\Delta\sigma$ Stress in kilopounds per square inch opening the crack
8. $N = (8.3 * 10^8/\Delta\sigma^{3.25}) * (1/a_0^{0.625} - 1/a_f^{0.625})$
9. Surface crack length when through plate $= c = 6t/2$
10. Surface crack length when start $= a_0 * 3 * t$
11. Tension cycles each year $=$ Bales/day $*$ time in minutes/bale$*$ rpm $* 365 *$ S.F.
12. Years for N cycles or go through plate $= N/$Tensile cycle per year.

For the example shown in Section 10.2:
$\Delta\sigma = 2.5$ ksi, $a_0 = 0.0625$ in., $a_f = 0.5$ in., $t = 0.5$ in., rpm $= 49$, S.F. $= 1.0$, 300 bales/day.

A service factor of 1 means it operates 100% of the year. Time per bale = 1 min. Calculated expected life years = 29 years' operating full time.

With the life calculations done and the results sounding reasonable from actual life data and with the equations now on a spreadsheet some time travel is possible. What I mean is that variables can be changed on a new chopper machine to see how this compares with the life of 29 years. This is called a sensitivity analysis.

- Doubling the bales processed per day reduces the life by 50%.
- Doubling only the cyclic stress reduces the life by 90%.
- Doubling only the bale properties meaning the toughness and thus the shear stress reduces the life by 90%.
- Doubling the initial crack size reduces the life by 50%.

Life can be increased also if the unit isn't operated continuously; however, start-ups impose other loads not presented here.

Sensitivity type analyses are useful because they show the relative effect of a change. Even though the actual life was 20 years and not the calculated 29 years, doubling the bale properties will still reduce the life by 90% no matter what the actual life was.

10.4   DISCUSSION OF FAILURE AND POTENTIAL FIX OF PROCESSOR

This was a rather complex analysis in that it required several fairly sophisticated tools for analysis. Figure 10.7 illustrates the equipment and the failure crack.

In Section 10.2, Figure 10.6, the flexing loading is evident. In Figure 10.7, the support points are shown. The housing is welded directly to the feed end plate. What appear to be occurring is that the cutting forces, as they rotate, cause a bending moment (M) on this plate. One can picture a thin metal plate flexing or "oil canning" in and out as the moment rotates. This is the cause of fatigue discussed in Section 10.3. This flexing is shown in Figure 10.8, which is part of a finite element analysis end view of the end plate and its deflection pattern. The fixed sides are part of the feed hopper and the tunnel housing connects the feed hopper to the discharge hopper. The moment M is due to the bale chopping process and rotates clockwise with only one position shown.

The crack shape is also indicative of fatigue. It starts at a stress riser and then wanders because of the relatively low stress field in the plate due to flexing. The finite element analysis of the end plate verified this.

There are many unknowns in this analysis such as the amount the cutting force influences the magnitude of the cyclic rotating moment, the dynamic material properties and any service history damage that may have caused a larger crack. The sensitivity analysis shows that the failure zone and crack growth is very sensitive to the cyclic stress. With a stress of 5 ksi, the life is reduced to 3 years. That is a major concern with cyclic stresses near welds. Changes in the bale properties meaning tougher or

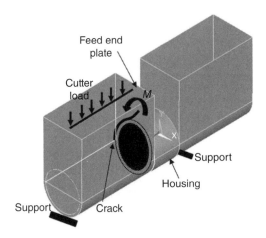

Figure 10.7   The machine and failure crack.

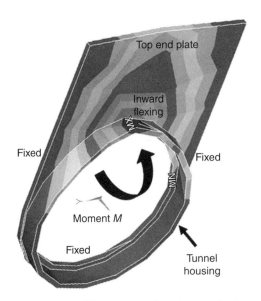

Figure 10.8   Side view end plate showing flexing.

colder material to shear could easily double the stress. This machine had actual years of operation of over 20 years before the crack was discovered. With a calculated life of 29 years being reasonable relative to the 20-year actual life, some justification of the model is possible especially if it agrees with other similar failures.

The difficulty in modeling the chopper is evident in Figure 10.9 as it is quite inaccessible when it's in service. This is when it is important to have the manufacturer's drawing with all necessary dimensions and operating procedures. The bales can be seen on the conveyor on the left being dropped into the chopper.

Figure 10.9   Chopper in operation.

Due to this stress/life sensitivity, strain gauge testing by the manufacturer in the area of the failure would be recommended before any modifications of the supports or end plates are performed. It is always wise to have the manufacturer agree in writing with the modifications the owner suggests. In this way, the modifications remain the responsibility of the manufacturer.

New machines purchased can have this written in the purchase specifications. For example, it may be stated that the cyclic stresses near welds should be less than 2,000 lb/in.^{2} and verified with strain gauge data.

This is the advantage of analytical modeling. Some very specific recommendations can be made. The manufacturer may disagree; however, it is now the manufacturer's responsibility to explain why this should not be a concern to the potential owner.

Options such as a larger capacity machine or a second machine could also be explored if production rates meaning bales per day processed are expected to increase.

As the reader can see, a concise and definite solution is not obtained from these type analytical models; however, sound suggestions can be made.

Sometimes, it is necessary to put a piece of equipment back into operation as soon as possible to minimize large production losses. This is when major decisions must be made with all interested parties involved. Should the machine be reinforced and welded in the areas analyzed, how long will the repair last, what are the alternatives, how long will it take to repair, or should additional testing be done are all valid questions an investigation team must address.

The analysis won't answer these questions, but it can help. Remember before the analysis, hardly anything was known about the failure. Now operating procedures,

Figure 10.10   Chopper model.

production rates, bale composition, and the effects of strengthening in the failed area are known to be important and need to be reviewed.

Section 9.2 shows why strengthening the flexing area by welding up cracks may not solve the problem and possibly shorten the equipment's life. This is because it will reduce the flexibility, possibly add defects, and also cause the metal to become brittle. It may however provide the owner enough time to purchase a new housing.

The one fact that made the results of the analysis reasonable was that other machines of a very similar type failed in a similar manner in about the same amount of time. A 20-year life is a respectable life for a welded structure machine with cyclic loading. As can be seen by the analysis it would not take a significant change in material properties, adverse operating conditions, or higher production rates to reduce the life. None of this was known before the analysis.

Since this is a book that gives away personal secrets, here is one that is used when building complex models to help visualize what is occurring. In the chopper model, a small model was constructed out of cardboard pieces found in the office and is shown in Figure 10.10.

By rotating the shaft as shown on the left, the blades forced the styrofoam piece onto the bottom of the box. When this was done, the end of the box was seen to flex inward where the failure occurred. This was because of the support points. Some other pieces in the model helped visualize other forces. While these type models are never shown to clients, they are helpful to the modeler especially when a finite element model illustrates the same effect. I remember when my children were small they always asked if they could play with me when I was building a structure out of dowels to investigate structural loads.

10.5   UNDERSTANDING THE SLOSHING EQUATION

10.5.1   Why It Is Important?

This may seem a strange topic to be considering in this book. The value is that it shows how there are many topics engineers may think they will never need to know but may be observed during their careers. When troubleshooting equipment, all types of phenomena may have to be explained to others and that's when an engineering background is extremely useful.

Tank trucks making sharp turns, ships carrying product, floating roof tanks being filled, or undergoing seismic activities [3] are examples of where sloshing can be a concern.

In this case, it was important to understand the surface wave motion in a large stirred and aerated processing tank to help identify the cause of damage to the internal horizontal deck.

10.5.2   What Is Sloshing?

Sloshing can be demonstrated by moving a cup of liquid on a table. The fluid will keep moving back and forth in a wave action after the cup motion has stopped. This is the liquid's fundamental natural frequency and is usually the most severe. Move the cup back and forth at this frequency and the wave will become larger. This base movement is called the excitation frequency and is similar to what occurs during earthquakes. The excitation could also be from mixers or liquids or gases pumped into the tank at critical velocities.

10.5.3   Developing a Simple Analytical Model

The wave motion is more complex than shown in Figure 10.11 when mixers, gases, or liquid filling are introduced, but the principle is similar. Visualizing a cup of coffee helped in developing this equation.

The analysis follows:

The unbalanced weight of the cylindrical wedge segment 1 is

$$W = \rho * \text{Volume} = \rho(D^2/12)\,x$$

The force balance to maintain equilibrium:

$$\Sigma F_x = 2W = \rho(D^2/6)\,x$$

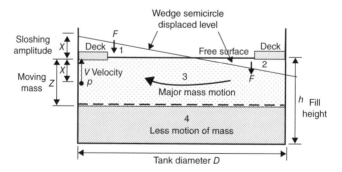

Figure 10.11   Development of sloshing equation in tank.

Recall $\Sigma F_x = m\ddot{x}$, where m is the mass accelerated meaning mass 3 and F are the forces acting for equilibrium.

$$(-\rho D^2/6)x = (\pi D^2/4g)z\rho\ddot{x}$$

where the thickness of the volume accelerated most is z.

Making the substitutions and simplifying,

$$(\pi z/2g)\ddot{x} + (1/3)x = 0$$

This differential equation is the familiar equation of motion for a single mass system whose solution is

$$f_n = (1/2\pi)(k/m)^{1/2}\,\text{Hz}$$

Making the substitutions results in the fundamental sloshing frequency for the tank:

$$f_n = (1/(2\pi))(2g/[3\pi z])^{1/2}\,\text{Hz}$$

The wave rises and falls a total of $2x$ at this frequency. The portion of the fluid in motion z is difficult to determine. However, comparing with experimental data, $z \approx D/16$.

Consider a tank 120 ft in diameter:

$$f_n = 0.15\,\text{Hz}$$

The time for the complete sloshing cycle is $t_s = 1/f_n = 6.7$ s. To move from p to the deck at a maximum velocity is one half this time.

Assume x is measured as 5 ft.

The maximum velocity is $V = x/(0.5t_s)$.

$$V = 1.5\,\text{ft/s}$$

This velocity can be used in the dynamic force of water equation [1, p. 144]:

$$F_{\text{deck}} = 1.93\,A_{\text{sqft}}V^2{}_{\text{ft/s}}$$

The horizontal internally supported deck has an area of 300 ft^2 in contact with the wave. The vertical force F_{deck} is calculated as 1,300 lb. A stress analysis using this cyclic force will determine if a fatigue failure was likely.

Observations of the actual wave action in this tank were chaotic and the excitation was not one obvious source. This quick analysis did show that cyclic wave action could cause significant fatigue loading of the deck.

The mathematical treatment of sloshing is complex [4]. The simple analysis shown here while not adequate for design is valuable for explaining sloshing to a failure analysis team and also for obtaining preliminary data.

10.6   FAILURE OF AGITATOR COUPLING BOLTS

It doesn't occur often but periodically the bolts on a multisection flanged agitator shaft can become loose. Once loose they usually fail in fatigue, which can result in significant damage to the gearbox and the rest of the system. Troubleshooting the cause can be difficult since there can be multiple causes such as

a. Bolts not tightened correctly
b. Shaft sections not correctly aligned
c. Wrong bolts or number used
d. Excessive shear or bending loads
e. Lateral or torsional vibration of system
f. Impact loading of agitator.

During my career, I have seen all of the above occur. A proper metallurgical analysis will provide valuable information on the cause. For example, it would determine if it was fatigue, shear, or a bending load. Torsional vibration of the system (*e*) would result in a shear type failure and a torsional analysis would be required.

Impact loading (*f*) of the agitator due to operational problems such as the level in the vessel being too low would allow the blades to go into and out of the product causing impacting dynamics. Also a periodic impact load on the agitator blades due to agglomerated product hitting the blades has occurred. Estimates on the magnitude of these types of problems can be determined by analytical analysis.

Failure due to bolt loosening causes the frictional clamping force of the bolts to be overcome and would require the bolt body to absorb the load in a shear mode as shown in Figure 10.12 [1, p. 44].

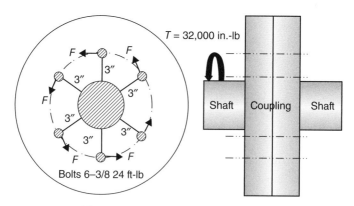

Figure 10.12   Torque/friction interface.

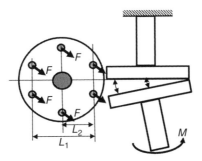

Figure 10.13   Flange "opening" load.

Another possibility is the flange clamping surface being "opened up" due to an impact load as in Figure 10.13. The moment M can be determined from agitator worse case horizontal loads. F represents the preload in the bolts.

Simply summing the bolt moments and setting equal to M for the case shown results in

$$F = M/[4(L_1 + L_2)]$$

This value of F should not be greater than the load the bolt is preloaded to. When it exceeds preload, the bolt will be stressed by the applied load and the flange will open. Once this occurs a failure can occur.

From agitator calculations [1, p. 31], the load trying to open the flange is

$F_h = 19,000 * HP/RPM * $ Diameter of impeller (D) in inches. With severe impacting, this value can be five times larger.

Consider $D = 96$ in., $HP = 100$, and $RPM = 30$ and the shaft length from the impeller blades to the coupling is 12 ft.

$$F_h = 660 \, lb \text{ so } M = 660(12) = 7,900 \, \text{ft-lb}$$

$$F = 7,900/[4(0.5 + 0.25)]$$

For this case, $F = 2,633$ lb.

The approximate load in a typical bolt would be

$$F_{bolt} = 60T_{ft\text{-}lb}/d_{in}$$

A 3/8 in. bolt load F_{bolt} when torqued to 24 ft-lb is 3,840 lb, so this is not a concern under normal conditions since F_{bolt} is greater than F.

Under impact, the load could be two times or more and opening and loosening could occur. High strength bolts with a higher preload would be a wise modification and would also increase the clamping force for the case shown in Figure 10.13.

10.7   CAUSES OF AUGER FEEDER SCREW FAILURES

10.7.1   What is an Auger Feed Screw?

Feed screws are used to transport granules and other products. They push or pull the product along. Figure 10.14 shows an auger type design. The shaft is a pipe with the flight spirals welded on. Several pipe sections may be welded or attached together.

The information shown here pertains to failures the author has analyzed. Failures can be quite complex and so more analysis than shown here may be required.

It's always wise to examine [1 p. 249] the failure zone to determine if it's a fatigue or impact weld failure. With such weld failures, the following calculations can be useful.

10.7.2   Concern with Torque Overloading

The torque required to move the product produces torsional stress on the screw pipe shaft. The nominal torsional shear stress S_{ST} due to the torque is a steady-state stress and if too high can cause a crack. It is maximum at the drive end and can be approximated as follows with units inches, pounds, and seconds:

$$S_{ST} = 320,856 \ \text{HP} * D_o/[(D_o^4 - D_i^4) * \text{RPM}] \ \text{lb/in.}^2$$

HP is the input horsepower, RPM the shaft speed, and D_o and D_i the outside and inside pipe diameters.

Starting up with product in the screw or "bumping" the drive motor to move stuck material can cause an impact. Experience has shown that this can result in weld

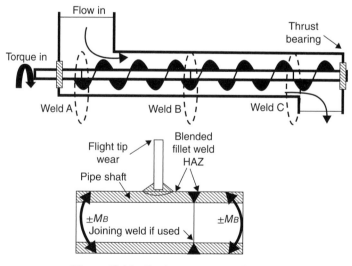

Figure 10.14   Feed screw auger and welds.

stresses of $3S_{ST}$ or more anywhere along the screw. Be concerned with weld cracking when this value exceeds 30,000 lb/in.^{2}.

10.7.3 Bending Stress Due to Screw Weight

The bending stress S_{bW} is due to the weight of the auger shaft assembly (W) and product with the pipe length L and results in sag. It is a cyclic stress maximum near mid-shaft (Weld B) when supported only by the end bearings and can result in fatigue failures. Failures can be due to the shaft diameter being too small. Mid-support bearings or a larger shaft may be required.

It is approximated as follows:

$$S_{bW} = \pm 1.3W * L * D_o/((D_o^4 - D_i^4)) \text{ lb/in.}^2$$

Welds have a low fatigue strength under $\pm$ cyclic stress due to inclusions and melt properties. The author's experience with equipment having in-service weld failures is that cracks grow when the cyclic bending stresses are over $\pm5,000$ lb/in.^{2}. Even good welds fail over $\pm10,000$ lb/in.^{2} [5]. The crack usually follows the weld in the heat affected zone (HAZ).

10.7.4 Bending Stress Due to Push-up

When the shaft is forced up δ by a hardened buildup or by large clumps the stress can be high. Every revolution will result in a bending stress. Wear on flight tips can be an indicator of this rub. An approximation of this cyclic stress $S_{b\delta}$, where $E = 29 \times 10^6$ lb/in.^{2} is

$$S_{b\delta} = \pm 8 * \delta * D_o * E/L^2 \text{ lb/in.}^2$$

Remember fatigue loading is cumulative so periodic overstresses due to inadequate cleaning or impacting can permanently reduce a weld's life.

10.7.5 Example of a Failure at Weld C Nondrive End Due to Buildup

This failure is interesting because it was in a low stress area; however, there was a hard buildup under the flight. Modifying the cleaning procedure helped eliminate failures by having $S_{b\delta}$ now equal to zero. The results of the analysis show the high bending stress caused by the measured buildup δ.

HP $= 5$, RPM $= 40$, 6 in. Sch 40 pipe, $L = 216$ in. $\delta = 0.25$ in.

$3S_{ST} = 1,400$ lb/in.^{2}; $S_{bW} = \pm\ 1,100$ lb/in.^{2}; $S_{b\delta} = \pm8,300$ lb/in.^{2}

10.8 TEMPERATURE OF A BLOCKED IN CENTRIFUGAL PUMP ON BYPASS

Over the years, I received several questions on how much time it would take for a pump to overheat when it is blocked in.

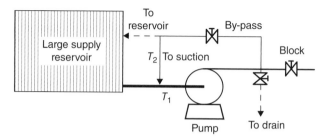

Figure 10.15   Blocked flow recirculation.

There are many reasons a centrifugal pump may have to be run on bypass. Maintenance checks of a new pump installation, start-up of a unit, and opening multiple flow lines manually as in irrigation systems are just a few. In all cases, the bypass flow should be enough to limit the temperature rise. Too much of a rise could result in reliability concerns.

The safest approach would be to contact the pump manufacturer and obtain test data on the minimum flow required for safe operation.

Many times the minimum flow data is not available and the following approach can be helpful for determining the temperature rise.

Let's discuss this with a simple example.

First consider Figure 10.15 with the blocked in flow being recirculated back to suction.

A simple analytical model using the specific heat equation is [1, p. 166]:

$$q_{BTU/min} = W_{lb}/t_{min} * C(T_2 - T_1)$$

From the centrifugal pump curve the blocked in horsepower HP_{BI} can be obtained and in terms of BTU/min:

$$q_{BTU/min} = 42.2 * (HP_{BI})$$

When curves aren't available, some use 40% of full load horsepower for HP_{BI}.

For this case, water is being pumped and the specific heat is $C = 1$ BTU/lb-°F. Also, 1 lb/min = 0.12 GPM.

Inserting these constants and rearranging the specific heat equation, we can determine the approximate temperature rise of this weight (W_{loop}) of fluid in a given time. It is the fluid weight of the case contents and recycle piping converted into gallons of the fluid in circulation (Gal_{loop}).

$$\Delta T \ (°F) = 5t_{min} HP_{BI}/Gal_{loop}.$$

This equation just determines the ΔT of the fluid in circulation with the blocked in horsepower over a given period of time.

As an example, assume we have a small centrifugal pump containing 15 gallons of water in the pump case and recycle line and it is recycled to suction. The pump blocked in horsepower from the performance curve is recorded as $HP_{BI} = 15$ and it is blocked in for 15 min. The temperature rise is $\Delta T = 75\,°F$ and since it is going to suction, the suction temperature and thus the pump contents will continue to rise. For example, if the temperature when the pump was blocked in was $80\,°F$, then after 15 min it will be approximately $155\,°F$ and after 30 min about $230\,°F$ and rising. This is for water and since the specific heat for many liquid hydrocarbon is about half that of water, the rise would be more with hydrocarbons. This could result in vaporization problems.

In Figure 10.15, the recycled water is also shown going to a large reservoir, to a drain or to suction. Going directly back to suction with no cooling is not recommended and would eventually cause the bulk fluid temperature in the loop to rise since the heat is not removed only transferred back to suction. To have no temperature rise, all of the heat generated would have to be removed by a cooler or eliminated from the system.

This has been based on considering only thermal problems with a blocked in pump. As a user moves away from the blocked in condition, other problems can arise if the minimum flow isn't adequate, such as vibration due to cavitation.

This discussion does not apply to positive displacement type pumps, which should never be blocked in or dangerous and damaging conditions will exist.

10.9   HEAT UP RATE AND RUBS ON A STEAM TURBINE

Simple analytical models have been used to troubleshoot equipment; however, sometimes a more detailed solution is required. This case history examines the heat up of a steam turbine and the thermal growth of the rotor assembly and the case. The FEM is used.

Figure 10.16 is an outline of a simplified model of a four-stage steam turbine. It's an axis-symmetric model, which means if you rotate the area about the Y axis, it will be a solid 3D model. This greatly simplifies the finite element modeling and is all that is needed for this problem. The results are easily explained to management.

A rotor disk had rubbed against a case diaphragm. One of the many potential causes the investigation team proposed was that since the insulation was left off the machine, the rotor shaft grew more than the cold case and touched for an instant. They requested an analysis be done to see if this was possible.

This axis-symmetric model contains the steam temperatures, film coefficients and calculates the thermal displacements of the machine as it heats up from the cold condition. This is called a transient coupled heat transfer stress problem. The nodal temperatures are determined from start-up and are automatically inputted into the stress model. The displacement is then calculated at any time period.

The purpose of the analysis is to determine if the growth difference between the rotor and the case at the rub point, meaning (b–a) in Figure 10.16 exceeds the cold assembly clearance. If it does, then there will be interference and a rub. Since a rub

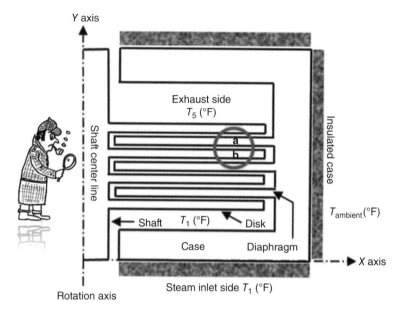

Figure 10.16 Axis-symmetric section of steam turbine.

never occurred with the insulation in place, the same rub point displacements will be compared with the insulation on and off.

The analysis indicated that even though the case was cooler without the insulation, so was the rotor assembly and the growth difference (b–a) was essentially the same. This is not unreasonable since the gas temperature at each stage stays about the same as it heats up the rotor and case. The rub problem was determined to be due to other causes, but the insulation was reinstalled for safety and efficiency reasons. An analysis was also performed on a 12-stage 40,000 HP steam turbine with similar results.

This shows that an analysis can be useful in determining what isn't the cause as well as what is the cause. By eliminating a possible cause, the investigation team can proceed in reviewing other potential areas of concern.

10.10 PNEUMATIC TESTING DANGERS AND BEWARE OF SAFE DISTANCES

Pneumatic testing of pressure vessels is usually considered when hydrotesting is not practical. For example, large vertical towers may have been hydrotested horizontally when they were built. Now that they are in service and vertical, the foundation may not be able to support the weight or the disposal of the contaminated hydrotest water, which is a problem. Many times project managers might say "what's wrong with pneumatic testing, the vessel is going to be pressurized in service with gas to a higher pressure than this anyway?" One answer to this is that the hydrotest is a much lower energy test. Much like pricking a balloon pressurized with air and one pressurized with water. One results in a bang and the other a slow leak and a whisper. Since the purpose of the test is to detect flaws, you don't want to find them with a loud bang but with a spray or leak. Remember air is compressible like a spring and water is not, so pressurized water results in much less stored energy. A vessel with a compressed gas contains many thousands of times the energy of the same vessel pressured with water.

It's an engineer's responsibility to question dangerous conditions such as when pneumatic testing is done instead of hydrotesting. Unfortunately, I still see failures [6, 7] and have been asked what a safe distance is to be at when performing a pneumatic test. I always try to provide other options instead of pneumatic testing and some of these are presented at the end of this section.

One major concern when defining safe distances from a pneumatic explosion are the many flying fragments and as we will see there is really no reasonable safe distance.

To illustrate the concern to a project team I once developed this analytical model. I didn't try to have them understand the model and only provided the results and options.

To understand the energy involved in pneumatic testing, the pressurized air was treated as a compressed spring. The energy released was used to propel a fragment horizontally. Here is a case where I visualized the pressure in the vessel as me pushing on a spring. The higher the pressure, the more the spring was compressed and the more energy was available. When the part failed, it would release all of this spring energy. Note that the analytical model could be for a vessel or a pipe with an end cap.

Figure 10.17 shows a fictitious gas spring in a vessel with some spring constant k in pounds per inch, compressed δ in inches. The diameter of the fragment is D in inches, its weight is W in pounds, the velocity it departs with is V in inches per second and the pressure p in pounds per square inch. Here δ could be the diameter of a vessel or the length of a pipe and represents the full compression of the fictitious spring.

The pneumatic spring constant is

$$k = p * (\pi/4)D^2/\delta \, \text{lb/in.}$$

The potential energy (PE) in a spring is (Section 9.6)

$$\text{PE} = {}^1\!/_2 \, k * \delta^2 = {}^1\!/_2 \, p * (\pi/4)D^2 * \delta \quad \text{in.-lb}$$

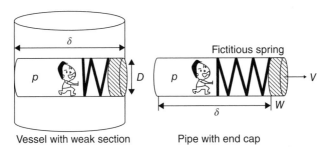

Vessel with weak section   Pipe with end cap

Figure 10.17   Compressed air model.

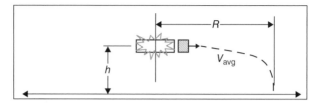

Figure 10.18   Fragment range.

The kinetic energy (KE) of the fragment W is

$$KE = \tfrac{1}{2}(W/g) * V^2$$

Equating $KE = PE$ and solving for velocity V of the fragment

$$V = 17.4D(p * \delta/W)^{1/2} \, in./s$$

This velocity can now be used in a simple trajectory calculation to see how far the fragment will travel horizontally.

R is the horizontal range that W travels at initial velocity V from a height h in feet.

$$R = (V/12) * (2 * h/32.2)^{1/2} \, ft$$

Figure 10.18 represents a fragment's flight path with some average velocity.

Here is the major problem with trying to define a safe distance as fragments of different sizes and weights will have different ranges. Stating a safe distance isn't possible unless you know the size. For example, with a 200 lb/in.^{2} pneumatic explosion of a pipe with $D = 10$ in., $h = 10$ ft, and $\delta = 50$ in., a fragment $W = 0.1$ lb will have $R = 3,600$ ft. However, one with $W = 25$ lb the range will only be $R = 228$ ft. You can see all the possibilities there are when the variables D, δ, W, and h are unknown as they will have to be estimated. Estimates will therefore be questioned if a safety incident occurs and will not be able to be justified in a court of law.

Detailed trajectory calculations that consider departure angles and air friction still require these variables to be defined and the range will again be an estimate.

Therefore, it's always prudent to look for alternatives to pneumatic testing of large volumes, recognizing that there may not be a reasonable safe distance.

For these reasons, always provide options for testing and don't try to "scare" the team into accepting your calculations. Here are some options for a project team to review:

1. Consider using localized hydrotesting instead of pneumatic testing and use small volumes, meaning sections of pipe instead of the whole pipe.
2. Consider nondestructive inspection of the modified welds and critical areas instead of pneumatic testing.

3. Review the historical hydrotesting of the modified vessel and only hydrotest the portions that have been modified rather than pneumatic testing. This can require temporary weld on caps. Yes, the welds could cause defects, but these can be ground away and the area inspected.

Work to recognized codes if you must pneumatically test at low pressures for leak testing only, but remember that there may be no reasonable "safe distances" or "exclusion zone." The safe distance specified in the code might be referring to the blast pressure wave acting on buildings and not the flying fragments.

Here's something to consider when discussing pneumatic testing. A vessel with an air volume of $200\,ft^3$ at $50\,lb/in.^2$, a truck traveling at 75 MPH or 1 lb of TNT, all possess about $1,500,000\,ft$-lb of energy. Thus my desire is to provide other options when testing large volumes.

Now I always like to review the literature to see if I can find any cases to compare the analytical model data with. You will see this done in the example on containment also. There is very little data available on the size and distance parts traveled in a pneumatic explosion, but some data is better than none. Table 10.2 represents some comparison data.

TABLE 10.2 Comparison Range Data from Literature

Case Examined	p (lb/in.^{2}), W (lbs), D (in.), δ (in.), h (ft)	Range (ft) $R_{actual}/R_{calculated}$
Shanghai LNG Terminal [6]	1784, 50, 10, 500, 20	1100/2200
Not enough data W, D, δ estimates		
Small air compressor tank explosion	200, 2.0, 10, 20, 5	200/360
Safeties bypassed tank corroded		
Air cannon and pumpkin contest	90, 16, 10, 1200, 80	3800/2700
Temporary end cap pipeline test [8]	1000, 154, 11.8, 1181, 3	330/650

While the actual and calculated range data results don't compare well because of the scarcity of information, it does show that large travel ranges are possible and that pneumatic testing can be quite dangerous. I would never propose a safe distance from the unit being tested because I wouldn't want to be standing there unless it was a few miles away. Thus, always propose alternatives to pneumatic testing. Remember this and speak up when someone at a meeting says "100 feet away from the unit should be OK."

10.11 CONTAINMENT OF A WRECKED INTERNAL PART

Parts break inside of machines such as an impeller inside of a turbine. We have seen news reports of the disks coming through engine cowlings on commercial aircraft engines [9]. What we don't usually see are flywheels in cars or turbine wheels in turbochargers coming through their housings. I have analyzed these since containment

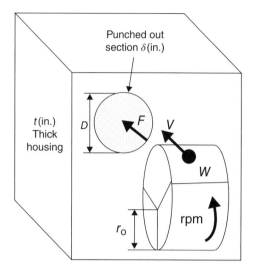

Punched out
section δ(in.)

t(in.)
Thick
housing

D

F

V

W

r_o

rpm

Figure 10.19   The containment model.

of parts is so important. You don't want them to fail at all but if they do, it's much preferable to have them demolish and remaining inside the housing rather than being launched through it. Unfortunately, this isn't always the case.

This analytical model is fairly straightforward but will show you the value of such an analysis and its limitations. It's general enough for a quick check to see if there is a concern. Observations on a turbocharger wreck were why this analysis was performed.

First let's sketch out an analytical model in Figure 10.19, which will describe what we want to do. We have a rotating disk, blades, or other debris pieces, which when they fail will have a departing velocity V. There is the weight W and shape of the part that has failed and there is the housing that must contain the broken piece W.

The problem is complicated since we have no knowledge on the size of the broken part. While rotors are known to fail in three or more sections, it might be prudent to consider half the weight of the disk as a start or possibly just a blade, if that's historically how failures have occurred. Normally for such a failure, some crack has started and grown and the disk fails at the bore first. Fatigue from many starts and stops and poor inspection at the recommended maintenance intervals are a major cause. Overspeeding of the disk can also cause these type failures. In clutch drives, the clutch assembly has been known to blow apart and go through the bell housing. That's why scatter shields or high strength steel housings are used in racing engines. Figure 10.20 is an example of such a failure.

In the model shown in Figure 10.19, the velocity $V = 2\pi * \text{rpm} * r_o/60$ in./s is the tangential operating speed, where r_o is the radius of the disk. It will be assumed the piece W moves tangentially in the direction of the motion shown until it impacts the housing. This has been observed on high-speed videos.

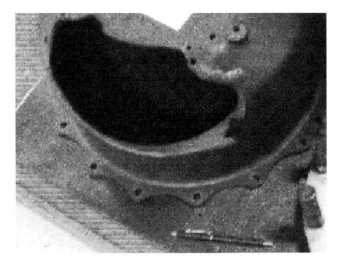

Figure 10.20   Exploded clutch bell-housing.

Since the sides of the housing are usually thick and not very wide, there's not much bending and mostly shear. The piece W will therefore be considered to be "punched out" in shear as shown on the shaded area in Figure 10.19. This is similar to what occurs in a metal punching operation.

The piece is punched out by being impacted by F and it will be assumed to have an area in shear of the perimeter of a circle of diameter D times the housing thickness t:

$$F = S_s * t * (\pi D) \text{ lb}$$

S_s is the ultimate shear strength of the material, which is obtainable as test data on many materials. The problem, of course, is determining what the diameter D is along with what W to use. Note that by using D to encircle the part, this doesn't mean the perimeter punch out shape is that of a circle. The shape could be anything encircled by its perimeter length πD. The penetration of the part W into the case is δ and is due to the velocity V. When δ is equal to t, the piece is considered to be "punched out" simply because it has gone through the housing thickness.

Equating the kinetic energy to the potential energy and solving for δ result in

$$\delta = (W * V^2)/(2 * 386 * F) \text{ in. where } F = S_s * t * (\pi D)$$

As long as δ is less than the housing thickness t, for this analysis, the disk piece is considered to be contained in the housing meaning we can define this as

Containment ratio $= \delta/t \ll 1.0$ and it is contained.

Consider where a turbine wheel came through the housing. For this example, it was assumed $D = 2r_0$ and $W = 1/2$ the weight of the rotor:

$$W = 3 \text{ lb}, \quad D = 2r_0 \text{ in.}, \quad \text{rpm} = 15,000, \quad r_0 = 4.5 \text{ in.}, \quad t = 0.375 \text{ in.},$$
$$S_s = 50,000 \text{ lb/in.}^2$$

The calculated containment ratio = 1.0.

It appears that there should be concern since the analytical model indicates there would be a shear out since δ is equal to t and indeed it had. It was also not clear if the parts had bunched up and the kinetic energy of the remaining rotating parts had caused much of the housing damage. In any event, it is now known that there is a good probability that it can exit the housing.

Many assumptions have been made in the development of the model and all can be debated such as

- The volume of diameter D is punched out in shear.
- W is estimated and so is D.
- All kinetic energy is converted into potential energy.
- The disk piece exits as shown in Figure 10.19.
- There is no wedging of pieces forced against the housing.

This is the problem with a simple analysis since they tend to require more assumptions. Normally, I try to justify the many assumptions by obtaining failure data from known failures and comparing the calculated values to them. In this case, failed impellers and projectiles will be compared to the containment ratio of the model in Table 10.3.

From this failure data, the model appears to be reasonable as it has predicted most of the noncontainment cases in the literature results. When you realize that there was no idea if a piece would go through before the model was built this is an accomplishment.

I only use this type analytical modeling to help determine the cause of a failure. I don't use them on new designs to predict if a design is safe or not. There are just too many unknowns. I use them for my own knowledge and to justify making the following type statements to the investigation team when a containment failure occurs:

- On high-speed or large gas turbines containment failures I would recommend detailed rotor inspections at specific life cycles in a manner recommended by the manufacturer. Containment is probably not possible due to weight restrictions.
- On small high-speed turbochargers and clutch assemblies, scatter shields, thicker housings, or containment blankets might be suggested.
- On a new rotor disk design, crack growth analysis should be performed to determine the life cycle inspection periods. I would also ask the manufacturer what type containment calculations they had performed. Usually this type information is very difficult to obtain [9]. When they say one test to destruction was performed and the blades were contained, my concern would be that on the next test they may not be. With complete engine blade off turbine tests costing millions of dollars more than one test is unlikely. It would be very valuable data if backed up with a sophisticated calculation method.

TABLE 10.3 Containment Verification Data

Item	Weight (lb)	V (in./s)	r_o (in.)	D of Perimeter Used πD (in.)	t (in.)	S_s	Calculated/Containment Ratio	Literature Result
Burst racing flywheel	7	2,500	3	6	0.25	50,000 Steel	0.97	Broke through housing
Helicopter Turbine rotor	2.7	9,000	4.3	8.6	0.3 titanium rods	80,000 Titanium	0.8 through combustion chamber	Contained in titanium rods and Kevlar
Burst racing flywheel	7	2,500	3	6	0.38	20,000 Aluminum	1.1	Broke through housing
Turbocharger impeller	3	7,000	4.5	9	0.5	50,000 Steel	1.4	Broke through housing much rubble inside
Projectile debris	0.008	30,000	Linear	0.22	0.3/0.5	50,000 Steel	3.0	Punched plug of D out $t = 0.3$
Projectile debris	0.008	30,000	Linear	0.22	0.3/0.5	50,000 Steel	0.5	Contained debris $t = 0.5$
Projectile debris	0.016	14,000	Linear	0.35	0.25	50,000 Steel	1.3	Contained

An unknown is what will happen inside the housing. A brittle failure due to other causes is shown in Figure 5.3. The broken pieces stacked up in the housing and while they were contained, they pushed the housing side out by the reciprocating mechanical forces. Most models, no matter how detailed, won't consider that mode of failure with any accuracy.

10.12   A CATASTROPHIC DISASTER

Earlier it was mentioned that just making repairs in kind or making minimum repairs can result in major consequences [10]. In the case of this disaster, there was a series of events that could have been avoided. It was primarily due to the culture established within the organization. There have been many such disasters and as mentioned earlier one only needs to perform a literature search on Catastrophic Disasters and many such cases will appear. Most have a similar pattern. The problem was known but not thought to be serious. Deadlines, budgets, and organization cultures controlled the outcome, not technical risks.

I bring the following case history into this book not because it could have prevented the catastrophic failure of the shuttle Columbia in 2003 but because of events that occurred. The shuttle was one of the most complex machines ever built with over 2.5 million parts and problems were bound to occur. However, organization problems on projects this big represent one of the quandary's engineers may face at some point in their careers. The Columbia Accident Investigation Board [11] stated that the failure cause was deep rooted in the culture of the NASA organization along with budget concerns. Similar statements were made in the Challenger accident [12] although for a different cause. The loss of the insulating foam that impacted the space craft was only part of the problem. Foam was lost from all of the shuttles. When something happens often without any consequence we become complacent. The thought is that it has happened before so let's not worry about it.

In the case of the space shuttle Columbia when the insulating foam came loose from the external fuel tank after it was launched, a large piece was known to have impacted the wing. Figure 10.21 was my simplistic sketch of this. At the time it was being mentioned by the news agencies that there did not appear to be any damage. What bothered me was the statement that was made on the effect. One report stated that it was like a styrofoam cooler mounted on top of your car. The report said, "Let's say it came loose and hit the car behind you. All it would do is break up into little pieces!" This view was strengthened since blurry photographic imagery after launch saw the impact and a cloud of possible debris at the impact point.

Now to a mechanical engineer this sounds absolutely absurd. I'm sure many engineers at NASA made this same simple calculation that I did. It's high school physics.

Like all engineering problems, one gathers data, makes a sketch of what the data says happens, and then does some calculations. For this problem, it is just using some of Newton's laws and calculating the deceleration of the foam piece. Videos were available that showed the foam coming loose and from these videos the approximate size could be determined. It was stated as "being bigger than the size and weight of

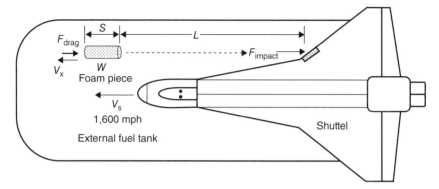

Figure 10.21 Impact of foam on shuttle.

a loaf of bread." So with a sketch of the shuttle and the meager data available some calculations were performed. Of course, the calculations would be very approximate and could have been off by quite a bit, but sometimes all we want is a rough approximation. After all, some were saying that the impact force was only a few pounds of force, merely a styrofoam cooler breaking up at 60 miles/h.

The following are the calculations I performed at the time while the Columbia was still in orbit. At that time, I was waiting to see similar calculations on the NASA site but never saw them.

In these calculations, I assumed that the foam piece hit the wing and the foam crushed its full length before it broke apart but didn't penetrate the leading edge. I was surprised not to see this simple analysis in the report [11]. I sure would have taken it to a meeting even to have someone tell me why I was wrong to be concerned or why it shouldn't be considered.

The first task was to understand with what velocity the foam hit the leading edge. Since it was important and interesting, here's my simplistic description. As soon as the foam pops off the external fuel tank its speed is the same as the shuttle. With no air resistance it would stay at approximately the shuttle speed. As soon as it gets into the airstream, and yes there is air resistance at 65,000 ft as the density is one-tenth that at sea level, the foam decelerates due to the drag force and its velocity slows down. The shuttle because of its constant speed then runs into the foam with an impact speed of the shuttle minus the foam speed.

Here's the approach used to determine the impact on the shuttle wing.

The drag force at 65,000 ft decelerating the foam will be considered as a constant force on the foam:

$$F_{drag} = 1/2 * C_d * \rho/g * A * V_s^2$$

Using Newton's second law the force needed to change the acceleration in distance L:

$$F_{accel} = m * a = (W/g) * (V_x^2 - V_s^2)/(2 * L)$$

Since the drag force is the force to decelerate the foam piece when it breaks loose:

$$-F_{drag} = F_{accel}$$

Solving for the velocity when the distance the foam travels from pop off to the wing impact zone is L in feet, then

$$V_x = (V_s^2 - F_{drag} * 32.2 * 2 * L/W)^{1/2}$$

Here A is the drag area in square feet and V_s is the shuttle speed in feet per second, W is the weight of the foam in pounds, and V_x in feet per second is the speed of the foam before it impacts the wing.

With $A = 1$, $V_s = 2400$, $W = 1$, $L = 80$ ft, $C_d = 1.0$, $\rho = 0.005$ lb/ft^3.

The impact velocity $V_{impact} = V_s - V_x = 530$ ft/s or 360 mile/h.

The following assumption Δt is the time the foam deforms when it impacts the wing, or $s = 1.5$ ft distance. Note no consideration is given to the foam penetrating the wing. In this analysis, the wing is considered rigid.

$$\Delta t = s/V_{impact} = 1.5/(530) = 0.0028\, s$$
$$F_{impact} = [W/g](V_{impact})/\Delta t$$
$$= [1/32.2](530)/0.0028 = 5,800\,lb$$

So with this we see that an approximation of the force with which the 1 lb. piece of foam would impact the wing was 5,800 lb or 3 ton. There's no need to do a stress analysis with this large of a force as it's quite evident that it will probably do a lot of damage on the thin carbon fiber wing. It's like dropping a truck on the wing. The force will be less because the following were not considered in the analysis:

- The foam will break up when it impacts distributing energy.
- The foam will penetrate the wing and so the full force will not be realized.
- The foam will partially glance off the wing reducing the contact force.
- The foam may do all of the above.
- Details on many other parameters were not available or estimated.

The analysis took about 8 h and most of the time was spent gathering any data that was available. A knowledgeable engineer would never have made a comparison of a styrofoam cooler. That's one of the problems with believing news reports that are trying to be dramatic, with no supporting facts. As an engineer working for the organization, you are told to say nothing because of the politics involved. In your career, you will probably have to make decisions on if you want to pursue certain issues and face the consequences.

As a comparison, using the methods just described, the impact force of a styrofoam cooler falling off a car going 60 miles/h and hitting one behind it going 60 miles/h and with sea-level air density, the force would be about 20 lb.

What it took to prove to the organization that there was a problem with the foam impact was a well-publicized test. An independent contractor fired a 1.7 lb piece of foam from a nitrogen pressurized cannon at 770 ft/s at a partial shuttle wing. It produced a force of 4,500 lb on the wing before penetrating it and putting a hole in it about 16 in. in diameter.

This one test did more than the most complex analysis could have done in convincing those in doubt. However, the analysis shows that even a thicker wing probably wouldn't have improved the situation. I was relieved when the shuttle program ended its 30-year mission in 2011 as the machine was just too complex to be managed as it was. Foam had broken off in later missions also but fortunately missed the wing.

For anyone interested in learning about the chain of events that can lead to disasters of this type, reading this report [11] is recommended. The one main point that really caught my attention was when the known foam failure problems were changed from being referred to as a safety issue to being called a maintenance problem. Not being regarded as a safety issue allowed the underlying cause never to be completely addressed.

Plant engineers witness this when hydrocarbon pump seal failures are maintenance headaches until a major fire occurs at which time they become safety concerns and the cause is permanently corrected.

Order of magnitude calculations to quantify speculation and verify computer program results are always useful and one engineers should utilize to make their point.

10.13 WHY ARE PARTS OUT OF TOLERANCE ON THE PRODUCTION LINE?

I've spent many years troubleshooting equipment but also time in research and manufacturing. In research, some rather sophisticated statistical design and analysis of data is frequently used when test data is available.

This section shows how very basic data analysis can be quite useful in troubleshooting. Most specialists are knowledgeable about plotting histograms and their value when they represent a distribution such as a normal distribution. As is said, "When looking for an answer, a good place to start is to follow the data."

In this case history, we review a production line that machines a precision groove in a casting bore for automotive use. Hundreds of thousands are machined a year and the groove out of roundness must be maintained within ±0.003 in. on the diameter since a seal fits in this groove.

Figure 10.22 shows one of the many clamping fixtures on the production line that hold the casting for machining. Periodically, a casting would be machined out of round and it was thought that the fixtures might be clamping too tightly or too loosely, thus causing the rejects. A load cell was designed to fit in place of the part to analyze the clamping force and was moved from fixture to fixture.

The fixture is designed so that when torque is applied via a hydraulic motor as shown, the part is clamped into place by screw action. A force F clamps the part securely in place. With too much clamping force, the part can distort and squeeze

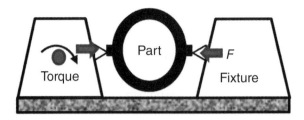

Figure 10.22   Clamping fixture.

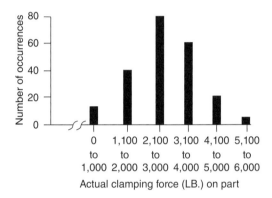

Figure 10.23   Clamping force on part.

into an oval shape. When the grooving tool makes the groove, it is a true circle but distorted when unclamped. This causes a rejected part.

A known torque of 300 in.-lb was applied to each fixture and the force measured on 215 fixtures. From the distribution shown, Figure 10.23 was plotted.

While there was much statistical analysis involved to determine the interaction of the many variables, one thing is obvious from this preliminary graphical display. The clamping force varies too much from fixture to fixture and since the force is directly related to the out of round of the groove, this contributed to the reject rate. The machining operation is designed for a 2,000–3,000 lb clamping force.

There were many causes for the rejects, but the worn-out fixtures and poor calibration of the torque motors contributed greatly. Making those repairs reduced the reject rate by 70%.

Here the spread of data represents a deviation from the norm and the question in troubleshooting should then be what has caused such a spread.

Note what the engineer had to do with his engineering background. He had to design and build a device, meaning the load cell clamping, procure and understand how to use the measuring equipment, understand how the fixtures worked, organize the testing plan, perform the testing, understand what the data was telling him and then to get with the plant management and personnel to design and implement a fix.

This job took about a month of the engineer's work year. The next job he worked on was totally different. Quite exciting and fulfilling!

10.14   FAILURES CAUSED BY AN IMPACT FORCE

When a deformed shape, a mass, and a velocity are known, an approximate determination of the force that caused this can be determined by equating the kinetic energy to the potential energy or

$$^1/_2(W/g) * V^2 = F_{avg} * S$$

Here is a case where a team investigating the crash of a company truck into a structure asked the author to determine the force this impact caused on the structure. It seems there was sensitive machinery on this platform and the data was required to determine if the machinery would need to be disassembled and inspected. This would be an expensive and time-consuming procedure.

Figure 10.24 is the analytical model considered and the speed limit in the area was 15 MPH (22 ft/s), with units of feet, pounds, and seconds.

The average force acting on the movable body of weight W is F.

$$F = {}^1/_2(W/g) * V^2/S$$

With a vehicle weight of $W = 4{,}500$ lb, a speed $V = 22$ ft/s, $g = 32.2$ ft/s^2 and a deformation of $S = 1$ ft, the impact force (F) on the structure is 34,000 lb. A side fact is that for the driver, the deceleration remains the same on the vehicle and driver ($W = 200$ lb) but only the weight seeing this deceleration is changed, so the force on the driver is $F = 1{,}500$ lb. It was a good thing he was wearing a seat belt.

The 34,000 lb force was then applied to the structure as an impact and the resulting load on the machine determined using a structural analysis program. Due to the design of the structure, the dynamic forces on the machines were negligible and they didn't have to be disassembled and operated without any adverse effect. This allowed the company to limit production losses as downtime was minimal.

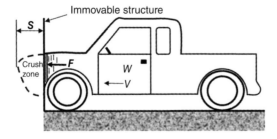

Figure 10.24   Deformation model.

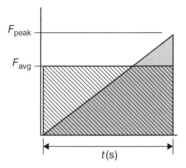

Figure 10.25   Relating average and peak force.

As a point of information, the peak force is about two times the average, but the average is usually used since it has a longer duration to do damage. Figure 10.25 shows this by equating the impulse areas $F_{avg} * t = 1/2F_{peak} * t$ so $F_{peak} = 2 * F_{avg}$.

After writing this section, I read a newspaper article that was discussing the deaths in high school football due to brain injuries thought to be due to helmet impacts. As a parent this was devastating to read and as an engineer you want to do something to stop this from happening. The first thought was recalling air bags in automobiles and wondering if something similar were possible for a football helmet besides the cushioning material now used. While helmets have come a long way, a little more cushioning might help. Possibly something like a cap that fits on the head under or over the helmet made of bubble wrap or something like it. As engineers a rough calculation can be performed to see if this is feasible using the method just used for the vehicle crash.

First a literature search to obtain some information.

- Average top speed of a player $V = 30$ ft/s
- Average weight of the brain $W_b = 2.5$ lb
- Average head hits per season 200–1800
- G forces = acceleration/gravity constant
  - Extreme football 150
  - Concussion 100
  - F-16 aircraft 9
  - Roller coaster 5

The average force on the brain for a concussion is therefore

$$F = W_b * G's = 2.5 * 100 = 250 \, lb.$$

But from the previous vehicle crash derivation $F = (1/64.4)W_b * V^2/S$.

Solving for the distance the brain moves meaning S:

$$S = 0.14\,\text{ft} \quad \text{or} \quad 1.68\,\text{in.}$$

I'm not sure what this distance (S) is or means but increasing it will reduce the average force. Maybe it's the brain in the brain fluid and its deflection or maybe the absorption design of the helmet. Probably all of these occur.

Let's say it's possible to add 1 in. of a bubble wrap type of a beanie cap to wear under or over the helmet.

G's $= 100 * 1.68/2.68 = 63$ or almost a 40% reduction of force on the brain. Since impacts are probably a cumulative event like metal fatigue, the reduction should help. How much it will help is unknown as is how to fit such a cap design under the helmet.

In any case, this illustrates how an analysis in one area can have the engineer thinking about using it elsewhere. Developing such a cap, performing testing, and trying to sell it would be a significant amount of work and expensive effort. The legal implications would be huge, so it's better left for others in the helmet industry to investigate.

10.15   DESIGN OF AN AIRCRAFT TOW

It was mentioned in Chapter 7 that the author had designed, built, marketed, and sold a device that would push small aircraft in and out of hangers. Figure 10.26 shows the towing operation pushing a 2,300 lb aircraft.

It was designed to be small and portable with no pushing effort for the person operating it. They would just hold the handle and steer the aircraft in or out. The market seemed to be happy with an electric gear-motor version that was easy to start in all weather, low in cost, lightweight, and maintenance free compared to gasoline engine-powered units. Technically, the product worked fine but as mentioned it was very expensive to advertise and market. All the design, welding, machining, and assembly was done by me and after building and selling several units a suitable shop to do the work, at a reasonable cost, could not be located. The cost they wanted was prohibitive for the selling price asked so production was stopped. It was just too much effort for the profit.

Some steps in the development of the unit were as follows:

1. Determine market size, competition, design and dimensions, and competitive price for the tow.
2. Determine required speed and push force of the tow.
3. Determine motor size and gearing to obtain speed and required push force.

During the design phase, calculations were needed to determine the force required to push the aircraft up a small grade. Since this is a book about engineering, the approach used will be shown. Someone without the necessary mathematical skills would have just tried something and if that didn't work would try something bigger,

Figure 10.26 Aircraft tow as sold.

such as gear motors, sprockets, or tire size. A friend who was the best fabricator I have known had this talent. Once when he was building a high-speed gearbox from just a billet of aluminum, I asked him how he knew the gears were the correct size and type. He said he had seen the same type gears used at twice the horsepower at the same ratio and speed in race cars. Not a bad design method and this was a brief insight into his design philosophy. He had a wealth of experience building and racing drag boats, top fuel dragsters, and aircraft thus his large database of information. He did it all meaning designing, machining, welding, testing, and assembling. Bud truly was a talented fellow.

In this analysis, the benefit of using the weight of the aircraft's forward momentum to help get the aircraft up a slope will be investigated.

The force F available to start to push the aircraft is

$$F = \mu_{\text{static}} * N = 0.9 * 100 = 90 \, \text{lb}$$

Here μ_{static} is the coefficient of friction between the tire and the pavement and N is the force pushing down on the tires. It's due to the force on the handle and also the weight of the tow and is determined by summing the forces and moments on the tow. F is then the force available to move the aircraft and the gear motor is sized to do this. With this design, the tires will slip on the pavement when over 90 lb push force

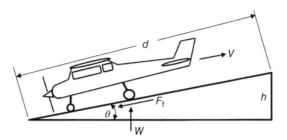

Figure 10.27   Aircraft coasting to a stop.

is developed with 5 lb on the handle. More force than this wasn't considered wise. Too much pushing force could damage the small attachment point connection to the aircraft. Slipping would be a safety fuse of sorts.

Usually, the slope to the door on some hangers to keep rain from entering is a 2% slope or about $1°$ and the tow must be able to push up this slope.

This is shown in Figure 10.27 with slope θ with a maximum height h.

$$F_{slope} = W * \sin \theta$$

Rolling friction is also present and that is the force required to just roll the aircraft on a surface due to the tire flexing and rolling resistance and must be added to this. With $\theta = 1°$, $W = 2{,}300$ lb, and $\mu_{rolling} = 0.015$, the force required to push it up the slope is

$$F_{slope} = W * \sin \theta + \mu_{rolling} * W = 74.5\,\text{lb}$$

and since the wheels will slip at 90 lb, the tow should push up this slope without the wheels slipping.

The tow was designed with a forward velocity of $V = 6$ ft/s. This results in a rolling momentum that can add more pushing force to move up even steeper slopes. Just like a car coasting up a hill if it loses power at the bottom, so extra momentum force will be available due to V and the mass of the aircraft. This will put no additional load on the connection to the aircraft.

The work due to the rolling friction of the tires and kinetic energy at the bottom of the slope due to the velocity of the mass will be converted into potential energy when it stops rolling at the top, meaning the velocity is zero. This is because the form of energy meaning kinetic is just converted into another form meaning potential. Figure 10.27 shows the nomenclature for the necessary trigonometry although the slope shown is greatly exaggerated.

Kinetic energy due to mass - work due to rolling friction up the slope = potential energy

$$d = h/\sin \theta, \quad F_f = \mu_{rolling} * W * \cos \theta$$

$$\tfrac{1}{2}W/gV^2 - (\mu_{rolling} * W * \cos \theta) * h/\sin \theta = W * h$$

After some mathematical gymnastics:

$$V = [64.4 * h[\mu_{\text{rolling}} * \cos\theta / \sin\theta + 1]]^{1/2} \, \text{ft/s}$$

This is the velocity the aircraft needs to achieve to coast up the slope to a height h before the velocity is zero. Doing this will require no push force by the tow for this extra momentum push force. It's like a car coasting up a hill when the engine stops at the bottom. As long as the velocity is equal or greater than this, it should be adequate to go to the top of the slope. There would be no need for this aircraft tow if someone were strong enough to get the aircraft rolling at this velocity. Most of us aren't, especially on a hot Texas day.

For a tow design with a velocity of $V = 6$ ft/s, $\mu_{\text{rolling}} = 0.015$, a slope of $\theta = 2°$ with $h = 0.08$ ft (1 in.) could be managed with no push force.

This was adequate for most hanger entries and shows the power of momentum in designs.

10.16   SHAFT FAILURES AND CRACK GROWTH

Troubleshooting shafting failures plays an important role in industry. Many types of machines can have these type failures and have been analyzed by the author. Extruder shafts, auger shafts, turbine shafts, crankshafts, rolling mill shafts, and many others have broken in half. Figure 10.28 shows a 10 ft extruder shaft that had suffered mid-barrel wear and failed in rotating bending fatigue.

The fear always is that if the cause is not determined and mitigated another costly failure may occur again. There are many reasons for shaft failures, but rotating bending type failures frequently are the cause and start from a defect of some type. A metallurgical report may mention this but probably won't tell you why it occurred or what to do to alleviate the cause. This case analyzes such a failure.

Figure 10.28   Extruder shaft bending fatigue.

10.16.1 Bending Failures That Grow Cracks

The growth of a crack is usually a fatigue phenomenon. Most metals have cracks especially if there are welds. Even grain boundaries in the metal itself are a type of crack. Cyclic tensile stresses can open the crack and if these stresses are high enough these cracks will grow each cycle they are opened. There is a limit at which the stress field is not high enough for the crack to propagate. This means it will not grow in a ductile material when the cyclic stress or crack size is below this value. While debatable, from failures I have witnessed the lower stress value seems to be around $\pm 2,000\,\mathrm{lb/in.}^2$.

During much of my early career, I have used traditional fatigue analysis approaches such as the modified Goodman diagram to see if the design was adequate in fatigue. When I started using fracture mechanics techniques to determine the stability of cracks in old pressure vessels, I realized that crack growth calculations could be used on ductile materials instead. Traditional fatigue methods aren't sufficient when there is a sizable preexisting crack. Traditional methods usually determine how long it takes to develop a crack. When a sizable crack already exists, crack growth calculations can be a very useful tool. So let's see how it can be used.

Consider Figure 10.29 that shows a hollow shaft with a sag. The sag could be due to the gravity weight of a large rotor between bearings or a horizontal mixer shaft with a weight load on it. With every revolution of the shaft Point A goes through a tensile stress and a compressive stress when it gets to Point B in this sag position. A sag is like holding a rubber hose and turning the ends. The hose stays in the sag position as it is rotated and the tensile stress will tend to grow the crack. Now if the shaft were in a permanently deformed and bowed condition such as occurs with a thermal bow, which is not shown, for every revolution of the shaft Point A will be a constant tensile stress even at Point B. It is therefore not a cyclic stress, so a crack will not see cyclic stress and won't grow.

Figure 10.29 also shows a crack in a shaft with a sag. The analysis of crack growth is based on fracture mechanics but since the discussion will be on ductile materials, the equations are simplified. Shown is what is known as a thumbnail crack because

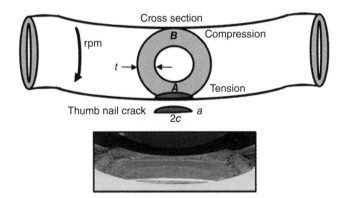

Figure 10.29 Sagging shaft with thumb nail crack.

of its shape. The depth "a" and width "$2c$" define the crack and is the typical nomenclature used in fracture mechanics.

The approach used will be to assume a small initial crack size and calculate the time it will take for the surface crack to break through the wall thickness (t). When this happens, the shaft will have lost much of its strength and a total failure will soon occur.

For this analysis, the initial crack depth (a) will be assumed as $a_o = 0.01t$ and the final at break through $a_f = t$.

The life equation for stainless steel is obtained by integrating the Paris equation [1, p. 340], which after making the substitutions is

$$N = [8.3 * 10^8 / \Delta\sigma^{3.25}] * [1/(0.01t)^{0.625} - 1/(t)^{0.625}] \text{ cycles}$$

here $\Delta\sigma$ in kilopounds per square inch is simply the cyclic tensile stress opening the crack and closing it. It will need to be determined fairly accurately by analysis or testing.

The time in years to reach this number of life cycles when the shaft rpm is known is

$$\text{Life in years} = N/(525,600 * \text{rpm})$$

10.16.2 An In-Service Failure Example

A failure occurred on a welded shaft made of a 6-in.-diameter pipe with a wall thickness of 0.55 in. rotating at 40 rpm. The shaft has a sag due to its weight and the tensile stress using beam theory with a distributed load is calculated as $\Delta\sigma = 8$ ksi. Proceeding through the analysis results in a life of 1.2 years. Since this was a horizontal mixer additional mid-span support bearings were required.

The thumbnail crack assumed in the analysis is $2c/a = 6$, which is fairly typical when looking at failures. This means the initial surface crack for this case was $c = 0.017$ in. and the final surface crack at breakthrough was 1.7 in. The stress would be higher because of the loss of the cross section, further accelerating the crack growth. A larger initial crack would produce a shorter life.

As a sensitivity analysis, assume the cyclic stress is reduced to $\Delta\sigma = 4$ ksi with the same initial crack. For this case, the life is 10.6 years and so a mid-bearing will certainly be a good solution.

10.16.3 The Assumptions and Comparisons

Like all simplistic models of crack growth, some justification due to actual experience is usually warranted. This type analysis would never be used for brittle fracture studies but is useful when it has predicted ductile crack growth with some success. So from previous case histories, Table 10.4 has been established.

Again the actual and predicted values vary considerably with my historical data; however, when no other data is available sometimes it's enough to make an educated decision.

TABLE 10.4   Various Crack Growth Cases

Case	Shaft	Cyclic Stress (ksi)	Initial/Final Crack size (in.)	Life Years Actual/Predicted
1. Extruder shaft (wear bending)	120 rpm	3	0.01/5.0	10.0/7.0
2. Dryer disk bending crack (measured)	10 rpm	6	1.0/2.0	0.4/0.4
3. Cracked gear tooth	650 rpm	16	0.1/0.4	Short/0.001
4. Vibrating conveyor welds	500 cpm	3	0.01/2.0	0.8/1.5
5. Vibrating conveyor welds repaired	500 cpm	1	0.01/2.0	+10.0 still going/54.0

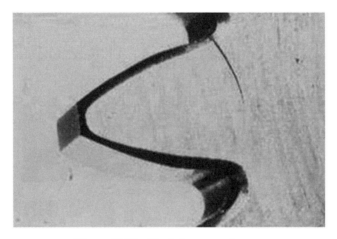

Figure 10.30   Crack in gear tooth root.

For example, when a crack was noticed on a gear tooth root and management wanted to know if it could be run until the next planned downtime in a week, an analysis was conducted. This is Case 3 and similar to Figure 10.30 although the crack was smaller.

With a predicted life of <8 h, I said it couldn't and explained the risk. If the tooth broke off, it could fall into the mesh and cause a major wreck. The repair would take a week rather than a day. This educated guess with some analysis is better than saying it's OK and hoping for the best. The risk just isn't worth the reward. There was no need to explain the analysis just to say that one had been done. Many times management is willing to take the risk for production reasons. This is fine as you would have presented the science as engineers should and their decision is being made for other reasons. As they say, "The ball is now in their court." Make sure you document your work and recommendation so your concern is known.

REFERENCES

1. Sofronas, A., Analytical Troubleshooting of Process Machinery and Pressure Vessels, John Wiley & Sons, 2006.
2. Spotts, M.F., Design of Machine Elements, 3rd edition, Prentice-Hall.
3. Nishi, H., Experimental Study of Floating Roof Integrity for Seismic Sloshing, 2008.
4. Ibrahim, R. A., Liquid Sloshing Dynamics: Theory and Applications, Cambridge University Press, 2005.
5. Pyttel, B., Gravenhof, P., Berger, C., Application of Fatigue Design of Welded Joints in Rotating Components, International Journal of Fatigue, 2012.
6. Pneumatic Test Explosion in Shanghai LNG Terminal, Chemical & Process Technology, March 2009.
7. Pneumatic Test Failure in Mississippi Pipeline Project, July 2009.
8. 12th Annual IPEIA Conference, Canada, 2008.
9. Mathis, J.A., Design Procedures and Analysis of Turbine Rotor Fragment Hazard Containment, DOT/FAA/AR-96/121, March 1997.
10. Bloch, K., Extreme Failure Analysis: Never Again a Repeat Failure, Hydrocarbon Processing Magazine, April 2009.
11. Columbia Accident Investigation Board, Report Volume 1, National Aeronautics and Space Administration, Government Printing Office; August 2003. 258 pp.
12. Report of the Presidential Commission on the Space Shuttle Challenger Accident (In compliance with Executive Order 12546 of February 3, 1986).

11

BENEFITS OF CONTINUING YOUR EDUCATION

11.1   BENEFITS OF AN ADVANCED DEGREE

By advanced degree, it is meant the master's or doctorate degree in the sciences. Why go on for your advanced degree? You can look at the pay scales and see that you will earn more during your career with an advanced degree. Some surveys say that relative to the Bachelors of Science it's approximately 20% more for Master's and 50% more for the Doctorate. It depends on many things but in general it's probably not too far off. You will realize an increase in opportunities, for example, teaching at the university level, along with the research world will now be readily available. Management in the research areas are also a realistic option. Research centers that deal with government agencies usually prefer to have professionals with advanced degrees on their roster.

One thing you should realize is that if you receive your advanced degree, especially your doctorate while you are working for the same company, they may not recognize it. They will see you are the same engineer doing similar work as before. This is a major reason I changed jobs after receiving my Doctorate. I went from design to research and was recognized for it in the new company.

Sometimes, you do things just to prove to yourself you can and for your own satisfaction and confidence. Those were my reasons for pursuing the Master's and Doctorate degrees. Every individual will have different reasons and I did appreciate the higher salary, college teaching, and research openings, which wouldn't have been possible to me without the degrees.

Survival Techniques for the Practicing Engineer, First Edition. Anthony Sofronas.
© 2016 John Wiley & Sons, Inc. Published 2016 by John Wiley & Sons, Inc.

I preferred to go to work after receiving my B.S.M.E. rather than continuing through to my Doctorate. There were two good reasons for this. I had just gotten married and wanted to settle down and go to work. The second reason was that I was out of money and had to go to work. This turned out to be a good decision because it gave me time to determine what courses to specialize in and what I wanted my thesis to be about. The best part is that if you receive your Master's while working, many companies will pay for your education.

This worked for my Master's but was tiring as I had to study and attend classes at night. I also had a family that needed attending too. So when I decided to go on for my Doctorate, I left my job and went full time. Now let me say that to do this you must have an understanding wife. It's a frightening thing to leave an established job of several years, with a steady income and a home, and take your family to live in another state, in an apartment with no income and just savings. The hope is that when you graduate you will find a job and it will all have been worth it. I can only imagine what was going through my wife's mind, but she did it with no complaints, just encouragement all the way.

Actually it worked out much better than either of us expected. My Doctorate was paid for with teaching fellowships, scholarships, and finally working full time as an engineer on a paid research project for a company. The project happened to be my doctorate thesis, which was directed by my Industrial Advisor. He was a brilliant man who also kept me on track for my degree. My salary was higher than before starting my Doctorate and my wife made many new friends. She went on to get her degree also probably to get even with me.

I had been hired by Bendix Research Laboratories as an engineer and developed the project to help solve a problem they were having in their manufacturing organization. I was given a budget and part of the project was experimental and the other was analytical. Responsibilities were for purchasing the programmable drilling machine, dynamometer, and instrumentation for gathering the data. After receiving my Doctorate, I stayed on as a Scientist with the Research Laboratory and then was Manager of Advanced Engineering for a couple of years. Figure 11.1 is a young and tired me with the laboratory equipment I purchased for obtaining data for the experimental side of my doctorate thesis.

11.2   IMPORTANCE ON SELECTING YOUR ACADEMIC ADVISOR

Advisors are the most important part of receiving your doctorate, since they help define your thesis and they provide their expertise and direction along the torturous path. My advice to colleagues who ask for it is that if you and your advisor don't get along well and you don't respect each other, find yourself another advisor. It can mean the difference between a 3-year program and a 10-year program. I was lucky enough to have two highly respected advisors. One was my academic advisor and the other was my industrial advisor. This was something unique at that time for The Doctor of Engineering degree (Eng. D).

Figure 11.1   Experimental equipment for data development.

The most important part of advanced degrees is the thesis. For the Master's, it is usually a scholarly dissertation improving on what has been done in the scientific world. A Doctorate thesis is to be new work that hasn't been done before and advances that particular area. It's very demanding and involves extensive research into an area along with the use of advanced analytical and testing techniques.

An initial defense of your thesis proposal, as a doctoral candidate, is when you are put before a group of your Professors, and you present your proposal. Questions on any aspect of the proposal or your course work to date are fair game. If they find major reasons that the proposal isn't adequate, they can make you change things. The worst-case scenario is that they find a major flaw and won't accept it. This means 1.5 years or more of your life have been wasted. This is every candidate's worst fear and one of the things a good academic advisor will prevent from happening.

My review had in it all of my professors in engineering mechanics, statistics, experimental design, computer programming, control theory, and other disciplines, and they were all extremely competent.

At the review, which my advisor chaired, an engineering mechanics professor asked me to write on the board, from memory, a rather long, complex equation, used in my proposal. This I did with some difficulty but to his and everyone else's satisfaction. I was relieved. He then proceeded to request that it be expanded to the next level of complexity, and then do a partial solution. My knees started shaking and the chalk kept slipping out of my sweaty hand. It could be done with enough time, but it was like having to finish a 3-h test in a minute. Then my advisor stood up, obviously sensing my hesitation and said, "I think Mr Sofronas has shown his knowledge of the equation to our satisfaction. There is no need to expand on it further. Next question please!" The rest went well as did the final defense and I received my Doctorate in record time because of his direction.

11.3   DIFFERENCE BETWEEN AN ENGINEER AND A SCIENTIST

Many times a subject is written about because someone has asked me for information on it.

One confusion that sometimes occurs is when people ask what is the difference between an engineer and a scientist. The answer is that at times each has to do the work of the other. When I was in a research laboratory, investigating new mechanical phenomena my title was scientist even though I was trained as an engineer. It just so happened that at that research facility those were the titles given for researchers with advanced degrees. In reality, this is the field of engineering science. Engineers frequently invoke aerodynamicist Dr Theodor von Karman's distinction between scientists and engineers: "The scientist seeks to understand what is, the engineer seeks to create what never was." For example, a scientist might envision and outline a machine such as a magnetic resonance imaging (MRI), but mechanical, electrical, and software engineers will be responsible for building and testing the first production models.

As for the mathematics, scientists tend to be more exacting and theoretical while engineers tend to use more approximations and empirical relationships to solve practical problems. Dr Albert Einstein was a scientist, actually a theoretical physicist, while the Wright Brothers with their inventions, experimentation, testing, calculations, and building of their aircraft, acted the mechanical engineering role, even though they didn't have a degree in the discipline.

11.4   BENEFITS OF CONTINUED EDUCATION

So much new is occurring in engineering every year that one would become quite obsolete if it weren't for continuing education. You can only achieve a certain level of expertise from on the job training. Your degreed learning stops when you graduate. There are many avenues to achieve this type education, some of which are listed as follows:

- Technical societies and trade seminars
- Technical publications
- Industry-sponsored seminars
- Company-sponsored seminars
- Review of recognized consultants work
- Online workshops
- Self-taught courses
- University courses
- Consultants

One of the most important areas I have used is consultants. Early in my career, I was doing what is called a torsional vibration analysis of ship systems. This is when you look at the dynamics of the internals of the engine, gearbox, clutches, propeller shaft, and propeller to see if there are damaging vibrations present. I had performed the analysis and then was told to have a well-respected expert perform the same analysis. My solution was then compared with his, which was invaluable for learning simplifying procedures. This was an expensive way for my company to train me, but it paid for itself because I analyzed the next four ship systems myself, with a high level of confidence.

The company sent me to attend a lecture at M.I.T. given by Dr Jacob Den Hartog [1], one of the greats in vibration analysis and author of many of the textbooks I had used.

He was older then and spent much time reminiscing on some of the unique problems he had solved during his career and that I loved. He spent about 15 min discussing a stability problem in exquisite simplicity and with much exuberance and animation. It was wonderful. A young engineer sitting next to me in a room full of 100 or more was unimpressed. He whispered to me that he thought this stuff was simple. I told him I had my Master's in this particular subject and never understood it clearly until then. I told the young engineer, "Enjoy it, you are in the presence of a great man!"

These type lectures are just a few of many I've sought out. It will be up to the engineer to find seminars of interest and educate management on why it is important for you to attend. In the case just described I was working on the vibration of ship systems for my company since they built the engine. Management felt that it would be beneficial for me to listen to an expert who had done many such systems and they were correct.

Learning in one's lifetime should never cease as long as you are able.

REFERENCE

1. Crandall, S.H., Biographical Memoir, (ed. Jacob Pieter Den Hartog), National Academies Press, 1995.

12

CLOSING GUIDANCE

12.1 DETERMINE WHAT YOU WANT TO ACHIEVE

Throughout our career, we will meet different personalities. How we relate to these different type people will certainly influence our success either positively or negatively.

Most engineers are similar to each other as are most managers when they are just entering the working world. Reference [1] has a list of 26 differences between engineers and executives. Here are a few of the more notable. Engineers tend to be introverts, methodical, like technical work, restrained, patient, and conservative. Executives tend to be extroverts, intuitive, like people, aggressive, impatient, and enterprising. The differences are important as they let each do their jobs well. It was difficult reviewing the list because obviously we all like to feel we have the characteristics of both and some might. Further review and being honest with myself showed me I was much better suited to be an engineer.

There are very few geniuses in this world and it's unlikely we will be working with them. People who are good at what they do weren't born with that talent, they grew into it. In other words, they learned it. We all went through approximately the same courses in school, be it science or management and some did better than others. The real test came when we applied what we had been taught.

True leaders don't do what is expected of them but more. True leaders not only try new ideas but also understand and manage the risks. True leaders are not afraid to make informed decisions. Note the key word "informed."

Survival Techniques for the Practicing Engineer, First Edition. Anthony Sofronas.
© 2016 John Wiley & Sons, Inc. Published 2016 by John Wiley & Sons, Inc.

12.2   MOST OF MY SUCCESS WAS DUE TO OTHERS

Much of my success was that I knew who the talented people were, utilized and learned from them, and tried to be helpful to others also. I was fortunate to have had excellent mentors to learn from during my career. It wasn't by chance because I chose them recognizing their unique abilities. Engineering also allowed me to limit risk in decision making by analyzing problems analytically by utilizing my "niche." Good managers do the same thing. They are successful because they choose good people to work with them, listen to them, and give them credit when it's due.

There is always someone who knows more about a subject than you. Like the bearing engineer mentioned in Chapter 10, it makes sense to use this type engineer when you have a unique problem you don't understand. There's just too much information to be an expert in everything, so utilize others when the need comes.

Technician, operators, machinists, and field hands have all humbled me with their equipment and machinery knowledge. When you work with them and get to know them they will share this with you. Talking with them during breaks I've talked with musicians, violin makers, big game hunters, drag car builders and racers, master machinists, church organ repairers, barbeque champions, Preachers, rodeo champions, and a list of others who were working hands at the company. People are interesting and likable. It's good to get to know them.

12.3   IT'S NOT SO MUCH WHAT YOU DO AS WHAT YOU HAVEN'T DONE

Of course, we have all heard the expression, "It's not what you know it's who you know." I guess I don't really agree with this in the engineering profession, because unlike some professions your accomplishments are most important.

However, I have noticed that for most successful engineers and managers, it's not so much of what you have done but what you haven't done. Having made several consecutive major errors in judgment or conflicts with management or the customer are far worse for one's career than having several technical successes to further your career. Most engineers and managers progress because they have a likable character, are good at what they do, and haven't made too many errors or generated too many enemies that can do them harm. This is especially true when it's time for a promotion.

What you do and how you do it is obviously very important. You can be a leader by taking on all types of jobs no matter how difficult or a follower by passing them off to others. Both will be noticed, but the leader will advance and the follower will probably stay relatively stagnant.

12.4   BECOME A MENTOR TO SOMEONE

During our career, most of us will be in a position to help some young people along in their careers. I had written this section to help others in this mentoring process utilizing some things I had learned and some I had heard from others.

Those of us who were fortunate to have a mentor in our careers know the value of this. As technical people, we should each be trying to direct others who are interested in the technical professions discussed here. It could be as simple as encouraging our children or others we meet on a career in the technical sector.

When the cost of a 4-year degree or the desire to obtain one is not present, we can explain that this is not the only path to success and show them that there are other excellent opportunities for a technical career [2].

We can also do this by utilizing our own career and experiences as examples along with those of others. Being a mentor and helping someone in their career is very gratifying.

A career is successful when it fulfills your needs and you have a passion for what you do. Monetary rewards and stability of employment are also highly desirable.

As an experienced engineer, I've come to notice the need for skilled engineers, technicians, and tradespersons in industry.

Over the years, the pool of experienced personnel has shrunk, mostly because of retirements, outsourcing, automation, lack of training by companies, loss of mentors in industry and primary education shortfalls, just to name a few. Vocational training in high schools has also faltered since many focus on 4-year college preparation [3].

12.4.1 Achieving Success Without a 4-Year Technical Degree

Not everyone is suited for success in college and there is a 40% drop-out rate before completing a 4-year degree. Over 60% of students now go longer than the 4 years. Fortunately, there are other excellent opportunities. I'm a proponent of trade schools for students who might not have been at the top of their high school class, don't enjoy school, or who want to work with their hands. These young people could be directed to trade schools to become craftsmen. Some may already have that ability. There are all types of trade schools for technical work such as machinists, welding, electrical, plumbing, heavy equipment repair, and others. These jobs can be high paying, are in need, and aren't easily outsourced overseas.

When I graduated from high school I enjoyed working on automobiles and machines. My parents directed me to a 2-year community college that provided a degree in Mechanical Power Technology with minimum expense. With this background in engine and component rebuilding, troubleshooting, machining, welding, drafting, mathematics, and other subjects, many satisfying jobs in industry were available. The analytical and design part was exciting and so I went on for my mechanical engineering degree and transferred half of the credit hours. These days selecting a 2-year program in Mechanical Engineering Technology will allow all of your credits to be accepted if you find the right engineering college and discuss this with them.

The point is that I'm sure I could have been quite successful with the background I had received even if I had not gone into engineering.

Two of my classmates wanted me to go in with them after graduation and open a car dealership. I said "No, I'm going to be an engineer!" Well, as they say hind sight is always 20–20 because they soon became millionaires since they were in the right place at the right time.

12.4.2 The Need for Technical Skills

At a recent meeting it was cited that the average experience level in a pipeline company employing 300 was 2 years. This can be rather concerning when significant skill training is necessary for critical positions.

A recent web search indicated 60 million articles on the need for various technical skill levels. Jobs are needed in the skilled trade area such as electricians, plumbers, heating and air-conditioning personnel, machinists, welders as well as degreed engineers, and technicians. There are technical schools and community colleges available which provide training for many careers. Make sure the school chosen is licensed or accredited by a nationally recognized agency. When they are not licensed or accredited and you work for a small company you may find it difficult if you decide to leave to look for another position.

In the oil and gas industries alone, it is estimated that one out of four current engineers, geoscientists, multiskilled maintenance professionals, process and production operators, health and safety professionals are currently eligible for retirement. This means there are many opportunities for skilled trades personnel. These industries need do fluctuate, so it's good to keep your skills current.

While "on-the-job" training might be possible without a formal education, the caution is what will happen if that job is no longer required. When your specialty is building aircraft wings and the job moves overseas, you may not have the skills to change jobs easily. This is where a licensed or accredited degree can help.

12.4.3 A 4-Year Engineering Degree Is Valuable

Besides the high salaries and the ability to always be able to find a job, it allows one to venture into areas not thought possible. This usually increases as one receives advanced degrees. During my career, I've been a machine designer, educator, author, researcher, project engineer, manager of advanced engineering, and worldwide technical problem solver on machinery and fixed equipment. After retirement, I still had my passion for engineering so I went into consulting, presenting seminars worldwide, writing technical books, and many articles. Not many careers will allow one that versatility.

When funding is a concern, a cost-effective 4-year engineering degree can be obtained by utilizing a local 2-year community college first, who have partnership transfer programs with a 4-year engineering university.

12.5 REMEMBERING THOSE BEFORE US

After 48 years working in the areas of mechanical engineering, I've learned an important fact. Everything we do is based on the work of others before us. In my case, engineers, mathematicians, and scientists much more talented than me.

I'd like to use this section to mention a few who influenced my career and others who contributed greatly to engineering mechanics.

In the world of mechanical engineering, the real first practical works were by Dr Stephan Timoshenko. Dr Timoshenko took the complex world of strength of materials and vibration to a new level. He used them to solve real-world problems and was gifted as a mathematician, engineer, and educator [4], and his books were my learning tools. Of course, there were others, but he stands out as a pioneer in the early twentieth century.

The second is Dr Jacob Den Hartog. Dr Den Hartog was trained as an electrical engineer, so his mathematical abilities were superb. Working with Dr Timoshenko his expertise developed in the area of vibration. World War II allowed him the opportunity to work many practical problems [5]. Like Timoshenko, he was known for his ability to solve more difficult problems using mathematical modeling.

His solutions were eloquent and clear. Den Hartog was also an excellent teacher. I once had the honor of taking a short summer course on vibration presented by him. It certainly was a memorable and exciting experience. I had been working on a difficult problem in the area of torsional vibration that quite frankly had me confused. In 10 min, he pointed out my erroneous thinking in a most polite way so as not to embarrass the then young engineer. One never forgets such kindness.

The books I've listed [6–9] certainly aren't the only ones written by these scholars but are the ones I have used the most.

Others have provided a great service to practicing engineers by compiling data, so it's available for us to use. I've always enjoyed Roark's Formulas for Stress and Strain [10] and used the fourth edition because of the format the equations were in. For back of the envelope calculations, it provided adequate solutions when used properly. And for design work where practical data was required, a copy of The Machinery's Handbook [11] was always at my side.

There were so many others and we use their works not even realizing it. Each new development comes from the combined contributions of others over many years. For example, Isaac Newton (1642–1727) and Gottfried Leibniz (1646–1716) were credited as the coinventors of calculus; however, the Greeks, especially Archimedes (287–212 BC), were using a calculus of sorts to determine the areas of shapes using limit geometry, which was a very rudimentary type of calculus. The Egyptians had methods to determine the compounding of interest as early as 1700 BC [12]. There is nothing completely new.

Galileo (1564–1642), Hooke (1635–1703), Newton (1642–1727), La Grange (1736–1813), D. Bernoulli (1700–1782), Euler (1702–1783), Poisson (1781–1840), Navier (1765–1836), and many other greats all played a part in experimental and analytical analysis with their contributions to mathematics to our modern analysis methods and were all mathematicians. In many cases, there were several great mathematicians in a family and you have to define carefully which one you are referring too. For example, there were 150 years of the distinguished and gifted Bernoulli mathematicians and it was Daniel who was into fluid dynamics along with many other fields.

Jean B.J. Fourier (1768–1830) was a French mathematician and physicist and while prolific in many areas he is best known in vibration for the Fourier series. Basically, this shows that the sum of periodic functions like sines and cosines and other functions can be used to decompose a complex curve into its harmonics. W.J. Rankine (1820–1872) is known to most mechanical engineers for his work in thermodynamics. What is little known is that Rankine also published work on the whirling of shafts [13] and he came up with the terms potential energy and critical speeds along with others.

So for all those before us who helped us on our career paths, let's not forget them for their superb work.

12.6   THOUGHTS ON THE FUTURE OF ENGINEERING

This book discusses engineering as it has applied to my career up to the time of this printing. As I reviewed the changes over time, I realized that the image of an engineer when I started in engineering is much different than now. Actually this is true for many professionals. In the future, things will be quite different.

Forty-eight years ago, space exploration, automobile design, new subsonic and supersonic aircraft design, computer usage, and many more technical areas were in their infancy and in need of all types of engineers. Engineering colleges could handle the new technologies in their programs and there were many job opportunities for the new-to-industry engineer. Companies were doing their own engineering and training their own engineers and outsourcing was not even imagined in those times. Many of the supervisors, managers, and even the vice presidents of engineering had engineering degrees. Those you worked for knew exactly what you were doing because they may have had your job. Most companies cared about developing engineers for long-term careers with the company and had in-house programs to do so. Senior engineers were expected to mentor engineers new to the organization and how well they did this went into their performance review.

Engineers of today will recognize that most of this has changed now and it will be quite different again when they are seasoned veterans.

Job stability probably won't be something engineers can reasonably expect in the future. Mechanical engineering is expected to grow at about 6–10% in the next 10 years. With the increase in globalization and the use of engineering services that are less costly in other countries, this will most likely continue and possibly even increase. Companies will do what they have to, to remain profitable. I have always said that if your work can be done less costly in another country watch out for your job. For example, large finite element analysis problems are very time consuming. This can be done at half the cost by outsourcing large jobs. This is work that use to be performed and valued by in-house engineers and companies.

The engineer of the future will need to excel in creative thinking and solutions of what needs to be analyzed as this is not easily outsourced. Most remote consultants won't have the unique relationship and understanding of your specialized plant equipment. You will need to do a simple analysis to be able to understand their complex computer-oriented solution. Even a simple in-house 2D FEA model or "back of the envelope" type calculation can be used to explain a complex 3D model. With this the engineer will know how to simplify the problem and can then easily explain the results to his or her management.

This means the need for an engineer to develop a niche, as discussed in Section 1.1, becomes even more important if one is to stand out. Versatility is very important and it is better not to have a niche that limits your expertise. Someone who was an expert at designing floppy disks wouldn't have many job opportunities these days. However, if the individual specialized in mechanical innovations, creative designs, and troubleshooting, the engineering world would remain wide open.

We go into engineering because we are innovative and creative individuals who want to make a difference doing important things. I enjoy knowing that I can walk through a plant and actually touch machines I have designed, have installed or determined why they had failed and modified them. Every time a certain diesel locomotive rolls by from a company I was employed by I can point to something on the locomotive that I had designed when I worked for that company. It may look slightly different, but it is still present.

Engineering uses our intellect and is challenging. We can look at history and see critical areas such as water supplies, modes of transportation, health technologies, and many other technical innovations engineers were responsible for and this is quite satisfying.

The problems of the future that require engineers are many. Addressing environmental concerns, increasing food production, reducing global warming, developing alternate energy production methods, future space exploration, and many more areas all will require engineers to conceive, design, and produce the new technologies to help solve these concerns.

An article stated that at one time there were only six engineers in the 112th United States Congress, out of 541 members. The majority listed themselves with law or political backgrounds. I was going to discuss how much better we might be with more engineers in Congress, until I realized it might be more productive having them work in industry. Engineers and scientists play a role in politics by being on task force teams and as scientific advisors. We as a group tend to be more subdued, working our magic in the background.

It's interesting to me to realize that much of what I had read and saw in science fiction comics and movies when I was young has come true; space travel, robots, two-way wrist TV and magnetic propulsion systems, death rays (lasers), trips to the bottom of the ocean in atomic powered vessels, and many more. I haven't yet witnessed time travel, antigravity devices, or teleportation designs, but I'm sure they are coming and no doubt engineers will play an important part in the equipment that utilize the technology. I imagine they won't be perfect either and will require engineers to troubleshoot the equipment.

At the writing of this book, there is an organization that was started in 2000 called Engineers Without Borders. The organization's projects impressed me and made me even more proud of my engineering degree. It shows other ways we help humanity in remote parts of the world through engineering.

REFERENCES

1. King, W.J., The Unwritten Laws of Engineering, A.S.M.E., New York, 1940.

2. U.S. Department of Education, Career Colleges and Technical Schools, Available on Internet.

3. Wright, J., America's Skilled Trades Dilemma: Shortages Loom as Most-in-Demand Group of Workers Ages, Forbes, 2013.

4. Timoshenko, S.P., As I Remember, D. Van Nostrand, 1968.

5. Crandall, S.H., Biographical Memoir, (ed. Jacob Pieter Den Hartog), National Academies Press, 1995.

6. Timoshenko, S., Strength of Materials, Part I, 3rd edition, D. Van Nostrand, 1955.

7. Timoshenko, S., Strength of Materials, Part II, 3rd edition, D. Van Nostrand, 1956.

8. Timoshenko, S., Vibration Problems in Engineering, 3rd edition, D. Van Nostrand, 1956.

9. Den Hartog, J.P., Mechanical Vibrations, 4th edition, McGraw-Hill, 1956.

10. Roark, R.J., Formulas for Stress and Strain, 4th edition, McGraw-Hill, 1965.

11. Horton, H.L., (Editor), Machinery's Handbook, 17th edition, Industrial Press, 1964.

12. Maor, E., e: The Story of a Number, Princeton University Press, 1994.

13. Rankine, W. J. On the Centrifugal Force of Rotating Shafts, Engineer, vol. 27, 1869.

INDEX

Survival Techniques for the Practicing Engineer, First Edition. Anthony Sofronas.
© 2016 John Wiley & Sons, Inc. Published 2016 by John Wiley & Sons, Inc.

Maneuvering through your career without any guidance has many potential dangers. Wouldn't it be wonderful if you had an experienced mentor to help you through critical decisions?

This is where *Survival Techniques for the Practicing Engineer* is unique. It presents the reader with actual rules, guidelines, procedures, and stories to help them navigate through their career.

This book will be useful for all engineers, managers, and technical personnel as it has many nuggets of information that will help ease them through difficult times. This is done with actual case histories and experiences and not vague generalities. It's like having your own mentor helping you along at every step with even those problematic decisions. All you have to do is turn to the appropriate section of the book.

Drawing on 48 years of industrial experience, the author has worked in education, research, management, technical design, manufacturing, and production in the petrochemical, transportation, and manufacturing industries.

This broad experience base makes the author uniquely qualified to talk on these subjects as they represent his personal experiences.

This is the type book the author wishes he had early in his career as he progressed through engineering.

Some of the topics covered are as follows:

- Developing your niche
- What to do and what not to do?
- Getting ahead in engineering
- The politics of engineering
- Successful problem solving and decision making
- Communicating effectively
- Becoming a consulting engineer
- Case histories on what a mechanical engineer does
- Advanced degrees and their benefits
- Benefits of continued education.